GW00535831

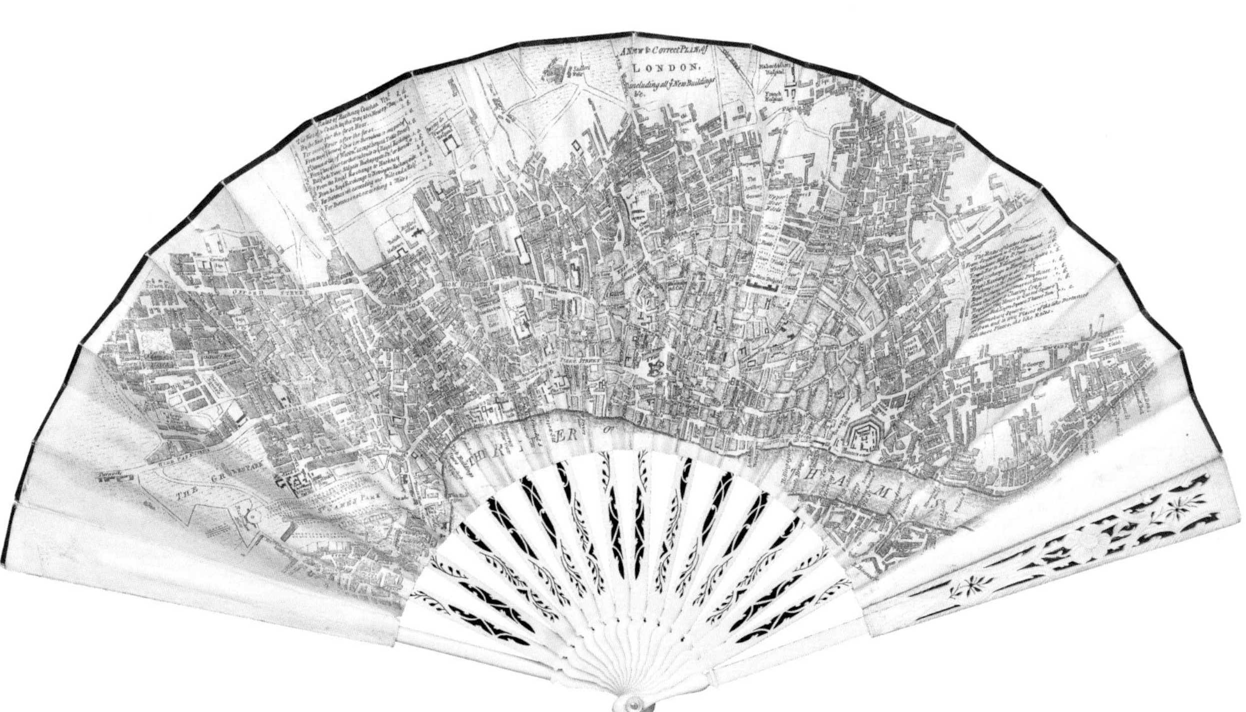
A New & Correct PLAN of
LONDON,
Including all y New Buildings
&c.

Daniel Crouch Rare Books LLP
4 Bury Street
St James's
London
SW1Y 6AB

+44 (0)20 7042 0240
info@crouchrarebooks.com
crouchrarebooks.com

ISBN 978 0 9567421 2 4

Catalogue edited by Daniel Crouch, Robin Hermanns and Nick Trimming
Design by William Joseph (williamjoseph.co.uk)
Photography by Louie Fasciolo and Ivone Chao

Cover: item 1; p3: item 48; p4: item 99

Terms and conditions: The condition of all books has been described. Each item may be assumed to be in good condition, unless otherwise stated. Dimensions are given height by width. All prices are nett and do not include postage and packing. Invoices will be rendered in £ sterling. The title of goods does not pass to the purchaser until the invoice is paid in full.

Printed by Park Communications on FSC® certified paper. Park is an EMAS certified company and its Environmental Management System is certified to ISO14001. 100% of the inks used are vegetable oil based, 95% of press chemicals are recycled for further use and, on average 99% of any waste associated with this production will be recycled. This document is printed on Chromomat, a paper containing 15% recycled fibre and 85% virgin fibre sourced from well-managed, responsible, FSC® certified forests. The pulp used in this product is bleached using both Elemental Chlorine Free (ECF) and Process Chlorine Free (PCF) methods.

Catalogue III
Mapping London

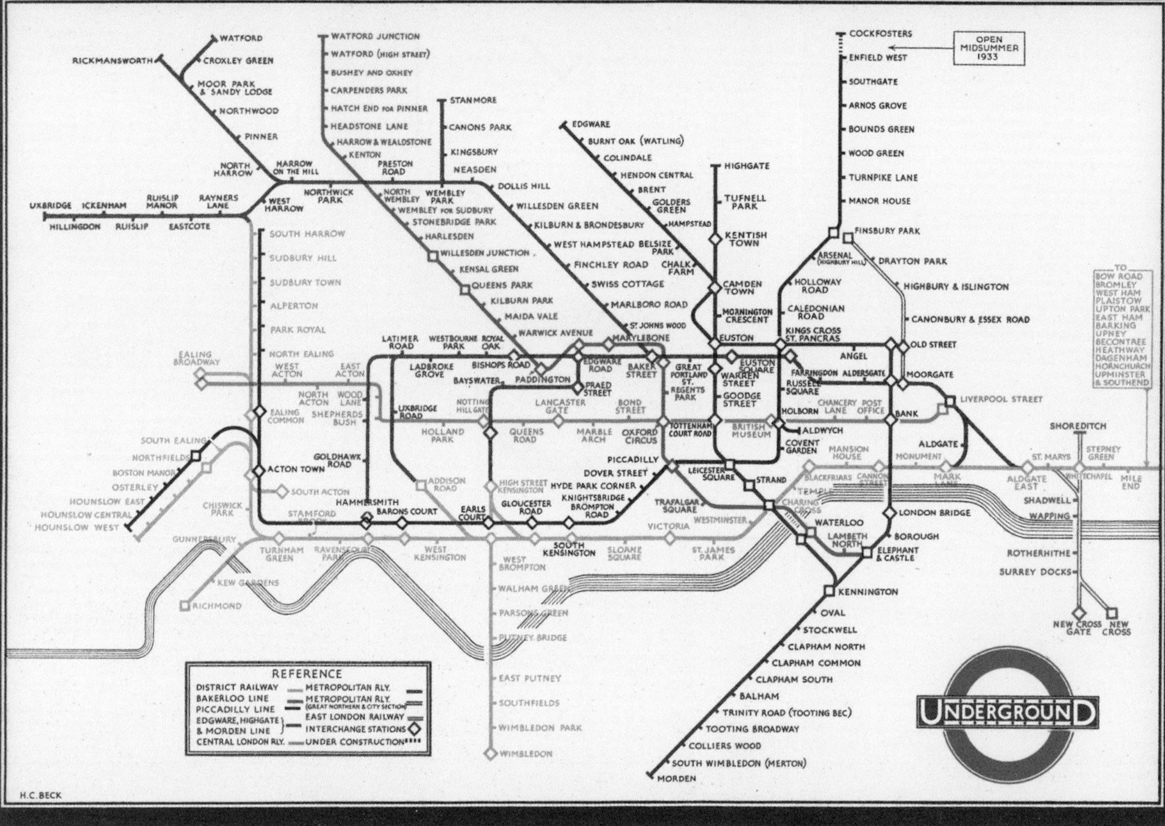
WATFORD
RICKMANSWORTH
CROXLEY GREEN
MOOR PARK & SANDY LODGE
NORTHWOOD
PINNER
NORTH HARROW
HARROW ON THE HILL
WEST HARROW
UXBRIDGE
ICKENHAM
RUISLIP MANOR
RAYNERS LANE
HILLINGDON
RUISLIP
EASTCOTE
WATFORD JUNCTION
WATFORD (HIGH STREET)
BUSHEY AND OXHEY
CARPENDERS PARK
HATCH END FOR PINNER
HEADSTONE LANE
HARROW & WEALDSTONE
KENTON
PRESTON ROAD
NORTHWICK PARK
NORTH WEMBLEY
WEMBLEY PARK
WEMBLEY FOR SUDBURY
STONEBRIDGE PARK
HARLESDEN
WILLESDEN JUNCTION
KENSAL GREEN
QUEENS PARK
KILBURN PARK
MAIDA VALE
WARWICK AVENUE
STANMORE
CANONS PARK
KINGSBURY
NEASDEN
DOLLIS HILL
WILLESDEN GREEN
KILBURN & BRONDESBURY
WEST HAMPSTEAD
FINCHLEY ROAD
SWISS COTTAGE
MARLBORO ROAD
ST. JOHNS WOOD
MARYLEBONE
EDGWARE
BURNT OAK (WATLING)
COLINDALE
HENDON CENTRAL
BRENT
GOLDERS GREEN
HAMPSTEAD
BELSIZE PARK
CHALK FARM
HIGHGATE
TUFNELL PARK
KENTISH TOWN
CAMDEN TOWN
MORNINGTON CRESCENT
EUSTON
COCKFOSTERS
OPEN MIDSUMMER 1933
ENFIELD WEST
SOUTHGATE
ARNOS GROVE
BOUNDS GREEN
WOOD GREEN
TURNPIKE LANE
MANOR HOUSE
FINSBURY PARK
ARSENAL (HIGHBURY HILL)
DRAYTON PARK
HOLLOWAY ROAD
HIGHBURY & ISLINGTON
CALEDONIAN ROAD
CANONBURY & ESSEX ROAD
KINGS CROSS ST. PANCRAS
OLD STREET
ANGEL
MOORGATE
TO BOW ROAD BROMLEY WEST HAM PLAISTOW UPTON PARK EAST HAM BARKING UPNEY BECONTREE HEATHWAY DAGENHAM HORNCHURCH UPMINSTER & SOUTHEND
SOUTH HARROW
SUDBURY HILL
SUDBURY TOWN
ALPERTON
PARK ROYAL
NORTH EALING
EALING BROADWAY
WEST ACTON
EAST ACTON
NORTH ACTON
WOOD LANE
EALING COMMON
SHEPHERDS BUSH
LATIMER ROAD
WESTBOURNE PARK
ROYAL OAK
LADBROKE GROVE
BISHOPS ROAD
BAYSWATER
PADDINGTON
EDGWARE ROAD
PRAED STREET
BAKER STREET
GREAT PORTLAND ST.
REGENTS PARK
WARREN STREET
EUSTON SQUARE
GOODGE STREET
FARRINGDON
ALDERSGATE
RUSSELL SQUARE
LIVERPOOL STREET
UXBRIDGE ROAD
NOTTING HILL GATE
LANCASTER GATE
BOND STREET
CHANCERY LANE
POST OFFICE
HOLBORN
BANK
HOLLAND PARK
QUEENS ROAD
MARBLE ARCH
OXFORD CIRCUS
TOTTENHAM COURT ROAD
BRITISH MUSEUM
ALDWYCH
COVENT GARDEN
ALDGATE
SHOREDITCH
SOUTH EALING
NORTHFIELDS
BOSTON MANOR
OSTERLEY
HOUNSLOW EAST
HOUNSLOW CENTRAL
HOUNSLOW WEST
ACTON TOWN
GOLDHAWK ROAD
SOUTH ACTON
ADDISON ROAD
HIGH STREET KENSINGTON
PICCADILLY
DOVER STREET
HYDE PARK CORNER
KNIGHTSBRIDGE
BROMPTON ROAD
LEICESTER SQUARE
STRAND
MANSION HOUSE
MONUMENT
ST. MARYS
STEPNEY GREEN
BLACKFRIARS
CANNON STREET
MARK LANE
ALDGATE EAST
WHITECHAPEL
MILE END
TEMPLE
CHARING CROSS
CHISWICK PARK
HAMMERSMITH
STAMFORD BROOK
BARONS COURT
EARLS COURT
GLOUCESTER ROAD
TRAFALGAR SQUARE
WESTMINSTER
VICTORIA
WATERLOO
LONDON BRIDGE
SHADWELL
WAPPING
GUNNERSBURY
TURNHAM GREEN
RAVENSCOURT PARK
WEST KENSINGTON
SOUTH KENSINGTON
SLOANE SQUARE
ST. JAMES PARK
LAMBETH NORTH
BOROUGH
ELEPHANT & CASTLE
ROTHERHITHE
SURREY DOCKS
KEW GARDENS
RICHMOND
WEST BROMPTON
WALHAM GREEN
PARSONS GREEN
PUTNEY BRIDGE
KENNINGTON
OVAL
STOCKWELL
CLAPHAM NORTH
CLAPHAM COMMON
CLAPHAM SOUTH
BALHAM
TRINITY ROAD (TOOTING BEC)
TOOTING BROADWAY
COLLIERS WOOD
SOUTH WIMBLEDON (MERTON)
MORDEN
NEW CROSS GATE
NEW CROSS
EAST PUTNEY
SOUTHFIELDS
WIMBLEDON PARK
WIMBLEDON
REFERENCE
DISTRICT RAILWAY
BAKERLOO LINE
PICCADILLY LINE
EDGWARE, HIGHGATE & MORDEN LINE
CENTRAL LONDON RLY.
METROPOLITAN RLY.
METROPOLITAN RLY. (GREAT NORTHERN & CITY SECTION)
EAST LONDON RAILWAY
INTERCHANGE STATIONS
UNDER CONSTRUCTION
UNDERGROUND
H.C. BECK

Introduction

"London is a modern Babylon."
Benjamin Disraeli

Welcome to the third catalogue from Daniel Crouch Rare Books. Within these pages you will find 100 plans showing London's rapid development, from the Tudors to the Windsors. You will see London the glutton, purged by fire, the home of the rich as well as the poor, and a refuge and opportunity for strangers; a city not dissimilar to the one we inhabit ourselves.

Foreigners, rich or poor, have always been a source of envy and fear, from the 'non-doms' in the City to the illegal immigrant or asylum seeker. Yet their influence upon London has been immeasurable, and is borne out by these plans. The Hanseatic League – arguably the first 'non-doms' – was responsible for commissioning one of the first maps of London in 1572 (item 1). Some 100 years later, another stranger, the great Czech engraver Wenceslaus Hollar, was working on Morgan and Ogilby's seminal plan of post-Fire London; the first true plan of the city (item 13). Finally, one must mention the French Huguenot John Rocque, who moved to London at the beginning of the eighteenth century to avoid religious persecution. His series of plans of London on one, four, eight, 16 and 24 sheets, (items 40, 41, 42, 43, 44, 46, 47, & 51) set new standards in the mapping of London.

Wealth features prominently in the mapping of the city: from Thomas Porter's depictions of the riverside palaces of the rich (item 9), to the, now lost, aristocratic town houses of the eighteenth century (items 16 & 36); one of particular note is Arlington House where the first cup of tea was supposedly brewed (item 22).The tea must have helped Londoners wash down the vast quantities of food they consumed, recorded on items 28, 29, & 30. Perhaps unsurprisingly, the presence of the poor featured less prominently on the early mapping of the capital, that is until Charles Booth's landmark 'Poverty Map' of 1882 (item 95), which recorded the distribution of wealth, with the most destitute described unflatteringly as 'vicious, semi-criminal'.

What all these plans have in common is the wish to render the chaotic city intelligible, in order that even a stranger could navigate London's labyrinthine streets; a wish that is repeated in many of the plans' titles. The greatest exponent of such clarity surely was Harry Beck, an engineering draftsman at the London Underground Signals Office, who created the famous tube map in 1933 (items 99 & 100).

In order to render this catalogue intelligible, we have listed the maps chronologically in terms of the development of the city – or, rather, cities – depicted, so that, for example, a map such as Vertue's 1737 re-engraving of Agas' plan of c.1560 appears as item 8 in the catalogue amongst the Elizabethan plans of London, as opposed to appearing around item 38 with images of the eighteenth century capital. Further, to aid comparison of a map's development, later states of maps appear adjacent to the earliest incarnation of that map included.

Daniel Crouch and Nick Trimming

The earliest extant plan of London

1 **BRAUN, Georg and Franz HOGENBERG**

Londinium Feracissmi Angliae Regni Metropolis.

Publication
Cologne, [1574].

Description
Double-page engraved plan, fine original hand-colour.

Dimensions
420 by 540mm (16.5 by 21.25 inches).

Scale
6½ inches to 1 statute mile.

References
Howgego 2 (2); Koeman 2433 state 4; in this state "Westmester" has been changed to "Westmunster" and the Royal Exchange has been inserted.

This magnificent plan was first published in Braun and Hogenberg's seminal town book 'Civitates Orbis Terrarum', 1572. London is depicted in birds-eye view from the south looking north. Above the plan is the title in Latin, flanked by the royal and the City of London's arms. In the foreground are four figures in traditional Tudor dress, together with two cartouches with text. The text on the left hand side is a paean to London, which is said to be "famed amongst many peoples for its commerce, adorned with houses and churches, distinguished by fortifications, famed for men of all arts and sciences, and lastly for its wealth in all things"; the text to the right deals with the Hanseatic League, which is praised for its global trade and its "tranquility and peace in public affairs", and names their trading hall in London, known as the Stillard.

Although published in 1572, the plan is clearly based upon information gathered some years earlier. St Paul's is shown with its spire, which was destroyed in 1561; the cross in St Botolph's Churchyard is shown, although it was destroyed in 1559; and York Place, so named in 1557, is given its old name 'Suffolke Place'. Upon the Thames, the royal barge can be seen, together with numerous ferrymen and sailing vessels. On the south bank of the river is the new district of Southwark, with its theatres, and bull and bear baiting pits. To the left is Westminster – connected to the City by a single road – with Westminster Abbey clearly visible. To the north of Westminster, cows are depicted grazing in open fields.

The view was most definitely derived from a 15-sheet city plan, of which only three plates have survived. The original plan was probably commissioned by the Hanseatic League, at sometime around 1550, hence the praise heaped upon the League in the text on the plan.

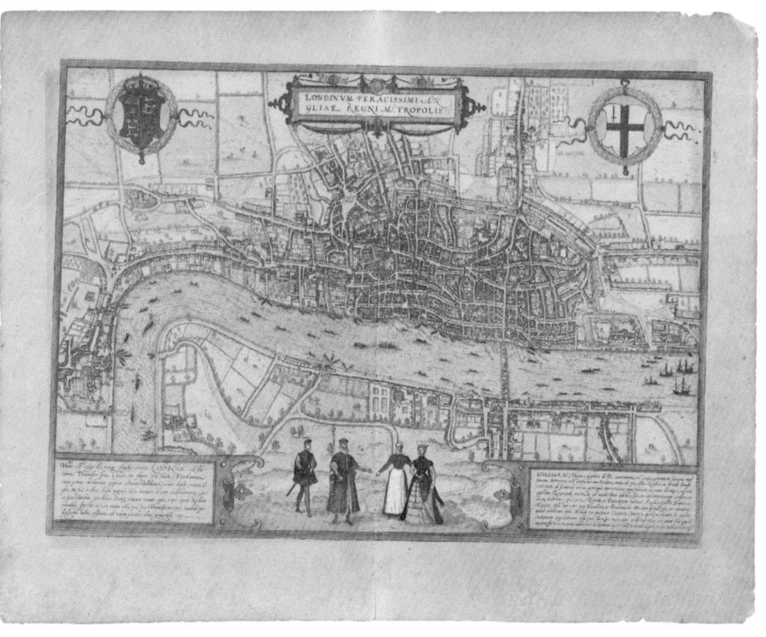

LONDINVM FERACISSIMI ANGLIAE REGNI METROPOLIS
Clarkenwell
Moor Gate
The Temple
Benams Castle
Thre Crane
The Bowll baytyng
The Bear bayting

Jansson's striking map of London

2 **[JANSSONIUS, Johannes]**

Londinum Vulgo London.

Publication
[Amsterdam, 1657].

Description
Double-page engraved map, blank on verso.

Dimensions
330 by 480mm (13 by 19 inches).

Scale
6½ inches to 1 statute mile.

References
Howgego 2 (3).

By the time Jansson published this plan in 1657, the London that it depicted was almost one hundred years out of date. The plan first appeared in various editions of Braun and Hogenberg's 'Civitates Orbis Terrarum' from 1572 until 1638. Soon after the final publication, Jansson acquired the plates and old paper stock. Although little revision was made to the plan itself, Jansson did remove the figures in the foreground, replacing them with a new title. The plan would be published in his very rare town book, 'Illustrorem principumque urbium septentrionalium Europae'.

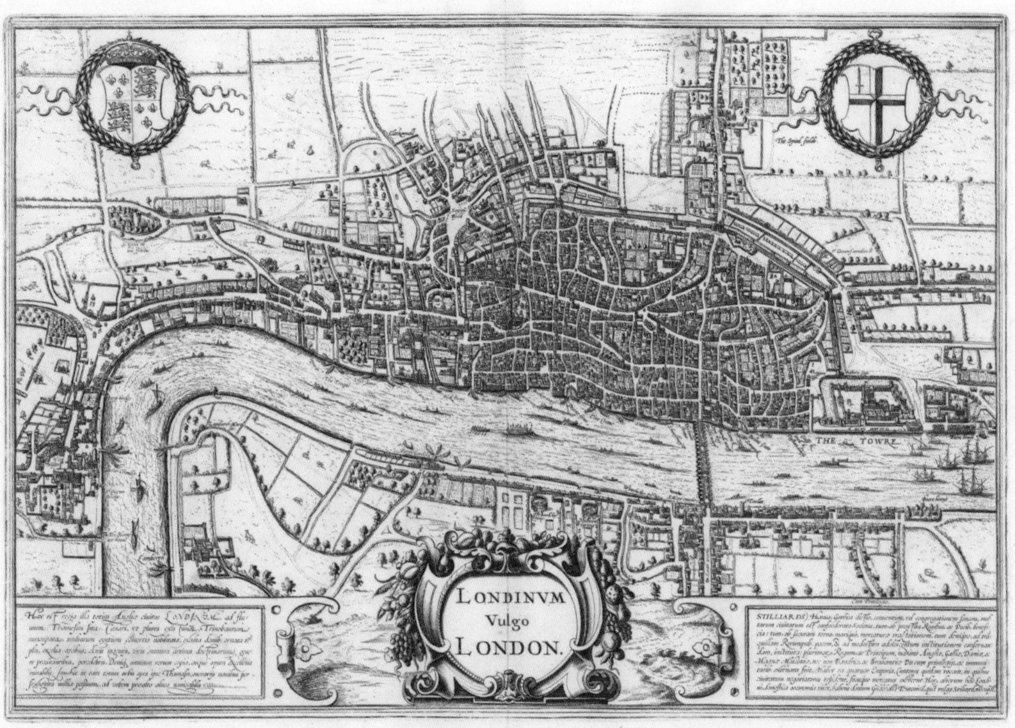

LONDINVM
Vulgo
LONDON.
THE TOWRE
STILLIARDS)

A woodcut

3 **BELLE FOREST,** Francois de

La Ville de Londres
Londinum Feracissimi Angliae
Regni Metropolis.

Publication
[Paris, 1575].

Description
Woodcut map, a few old folds strengthened.

Dimensions
300 by 480 mm (11.75 by 19 inches).

Scale
6½ inches to one statute mile.

References
Howgego 3.

Belle Forest based his plan upon Braun and Hogenberg's seminal map of 1572. The map depicts Elizabethan London in all its glory with bull and bear pits in evidence south of the river Thames, animals grazing in the nearby fields, and the river teeming with Tudor barges and sailing vessels. The only bridge across the Thames is London Bridge which joins the City of London to Southwark. The three cities of London, Westminster, and Southwark are depicted as quite separate entities at this time. In the foreground are four Londoners in traditional Tudor dress. The figures are flanked by text and, above the plan, the title is flanked by the royal and the City of London's coats-of-arms. Belle Forest's use of woodcut, rather than Hogenberg's copper engraving, gives the plan a subtler look.

The plan was published by Belle Forest in 'La Cosmographie Universelle de Toute le Monde' in 1575.

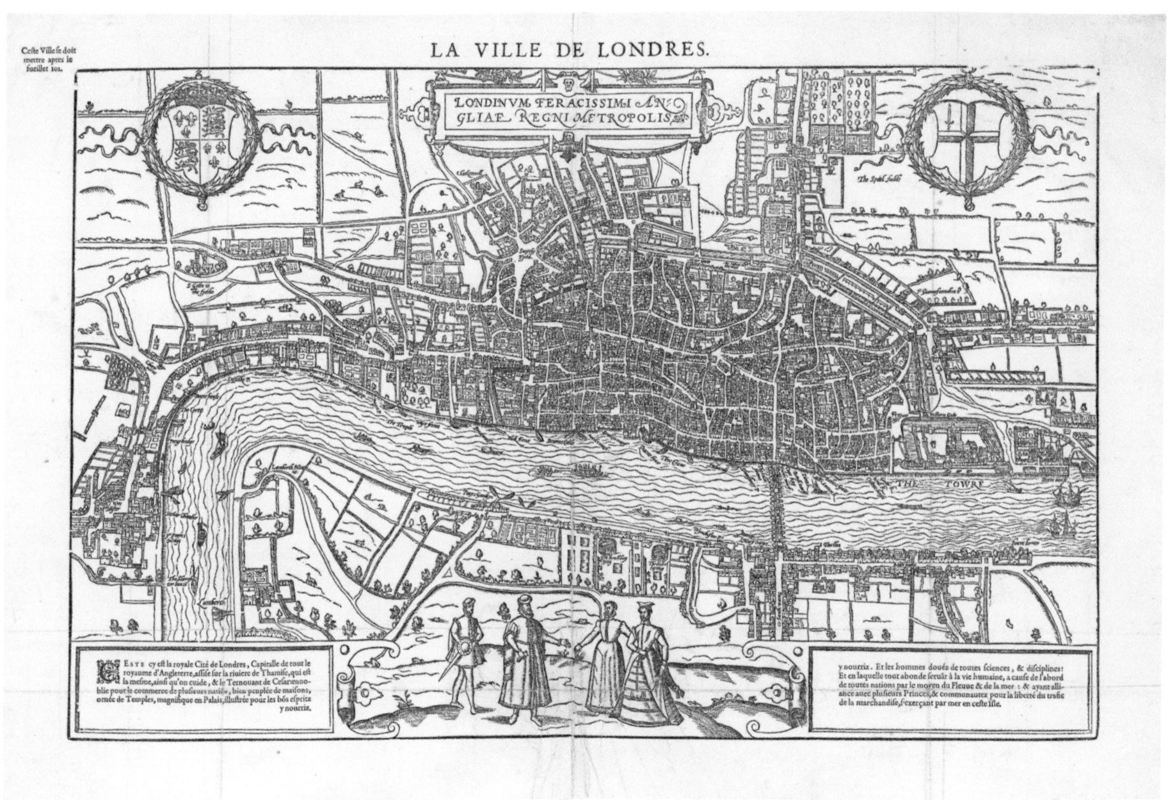

"One of the earliest map-views of London to have survived"

4 **VALEGIO, Francesco**

Londra.

Publication
[Venice, 1600].

Description
Woodcut map.

Dimensions
85 by 120mm (3.25 by 4.75 inches).

Scale
1¼ inches to 1 statute mile.

References
Howgego 5a.

The plan shows the City wall enclosing a somewhat haphazard collection of buildings, fields around it and a double line of buildings linking to Westminster. A line of unidentifiable buildings is shown along the south bank of the Thames. The map was probably based upon that of Braun and Hogenberg, and appeared in 'Raccolta di le Piu Illustre et Famose Citta di Tuttl il Mondo' (1660); and 'Unviersus Terrarum Orbis Scriptorum Calamo Delinatus' (1713).

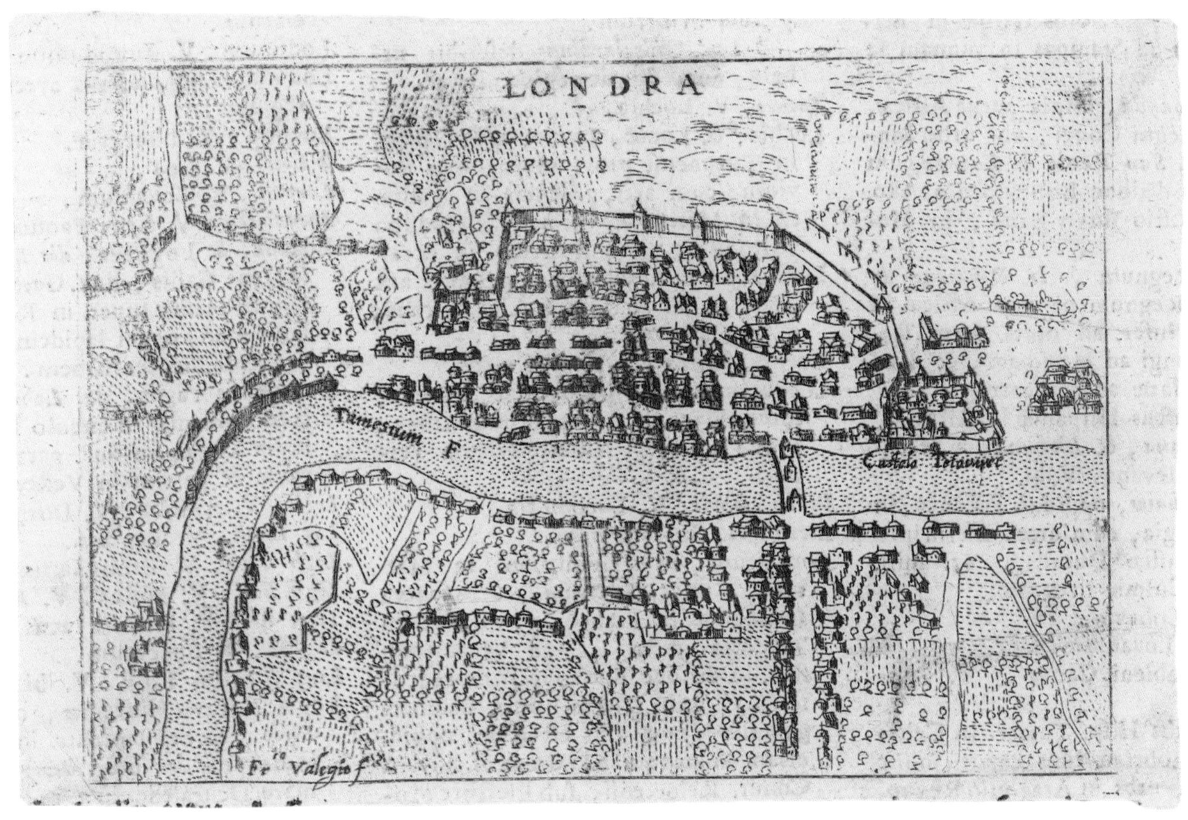

A plan from Sebastian Münster's 'Cosmographia'

5 **MÜNSTER, Sebastian**

Londen oder Lunden die Haupestatte in Engellande... Londinum Feracis: Ang. Met.

Publication
[Basel, 1598].

Description
Woodcut map.

Dimensions
222 by 355mm (8.75 by 14 inches).

Scale
5 inches to 1 statute mile.

References
Howgego 6 (1).

This map first appeared in the 1598 edition of Sebastian Münster's 'Cosmographia'. It is based upon Braun and Hogenberg's map of the city published in 1572. The title is on a swallow-tailed banderole above the map, and is flanked by the royal and City coats-of-arms. Below the map are four figures, copied from Hogenberg's original, flanked by two blocks of descriptive text.

Münster's 'Cosmographia', first published in 1544, was one of the earliest histories of the world. It was hugely successful, running into numerous editions thoughout the sixteenth and early seventeenth centuries in Latin, French, German, and English.

---

6 **MÜNSTER, Sebastian**

Londen oder Lunden die Haupestatte in Engellande...

Publication
[Basel, 1628].

Description
Woodcut map.

Dimensions
222 by 355mm (8.75 by 14 inches).

References
Howgego 6 (2); in this state the page nos. 84 and 85 have been substituted for lvii and lix.

Londen oder Lunden die Hauptstatt in Engellandt/am fluß Thamesis gelegen/auff das allerfleissigste nach jetziger Gelegenheit Abcontrafehtet.

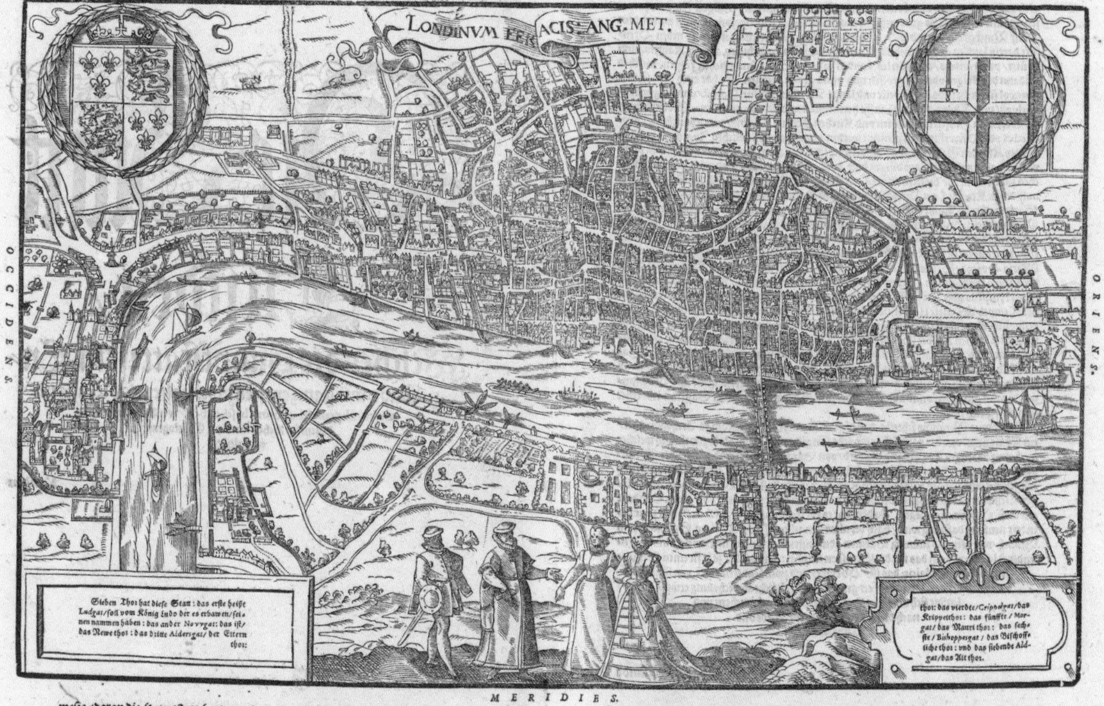

SEPTENTRIO.

LONDINVM FERACIS: ANG. MET.

OCCIDENS.

ORIENS.

Sieben Thor hat diese Statt: das erste heiße Ludgat/soll vom König Ludo der es erbawen/seinen nammen haben: das ander Novvgat: das ist/ das Newe thor: das dritte Aldersgat/der Eltern thor:

thor: das vierdte/Criphelgat/das Krippelthor: das fünffte/Morgat/das Mauri thor: das sechste/Bishoppesgat/das Bischoffliche thor: vnd das siebende Aldgat/das Alt thor.

MERIDIES.

mesis/daran die Hauptstatt Londen gelegẽ/die vor zeiten Trinouantum geheissen. Von dieser findt man also geschrieben.

Von der Hauptstatt Londen in Engellandt. Cap. xvij.

Es ist Londen ein gar alte Statt (deren auch Cornelius Tacitus in der Mittelsexischen Graffeschafft gedenckt) vnder den Stetten des gantzen Engelands die fruchtbarste vñ gesundeste/sechtzig tausent Schritt von dem Meere am Wasser Thamesis gelegen/hat in der breite ein vnd fünfftzig Grad vnd dreissig Minuten/in der lenge eilff Grad/vnnd zwentzig Minuten. Diese Statt wird in der alten Englischen Sprach genennt TroiNouant: lautet auff Latein Troia Noua: das ist New Troia: dieweil sie auß den Reliquien der Statt Troiæ zu den zeiten Bruti soll erbawen seyn/daher sie dann nur auß enderung etlicher Buchstaben von Cæsare (wie auch andere mehr meynen) Trinobantum genennt worden. Jhren Nammen nun betreffende/solle sie denselbigen von einem König/so Ludo geheissen/empfangen haben: vnd daß dem also/gebe dessen noch ein alt Thor der Statt gnugsame anzeigung/welches

er erbaw-

Speed apologizes…

7 **SPEED, John**

Midle-sex described with the most famous Cities of London and Westminster. Described by Iohn Norden, Augmented by I. Speed.

Publication
[London], Sold in Popes head alley against the Exchange by George Humble, 1611.

Description
Double-page engraved map, inset view of London, upper right, view of Westminster, upper left, St Peter's (Westminster Abbey), lower left, and St Paul's, lower right.

Dimensions
420 by 560mm (16.5 by 22 inches).

Scale
3½ inches to 1 statute mile.

References
Skelton 7; Howgego 7.

Map of Middlesex, published by John Speed in his 'Theatre of Great Britain' of 1611.

An open volume to the right of the map apologises for the lack of detail on the London plan and promises that a larger plan will be made. The two plans of London and Westminster were copied from those in Norden's 'Description of Middlesex', published in 1593.

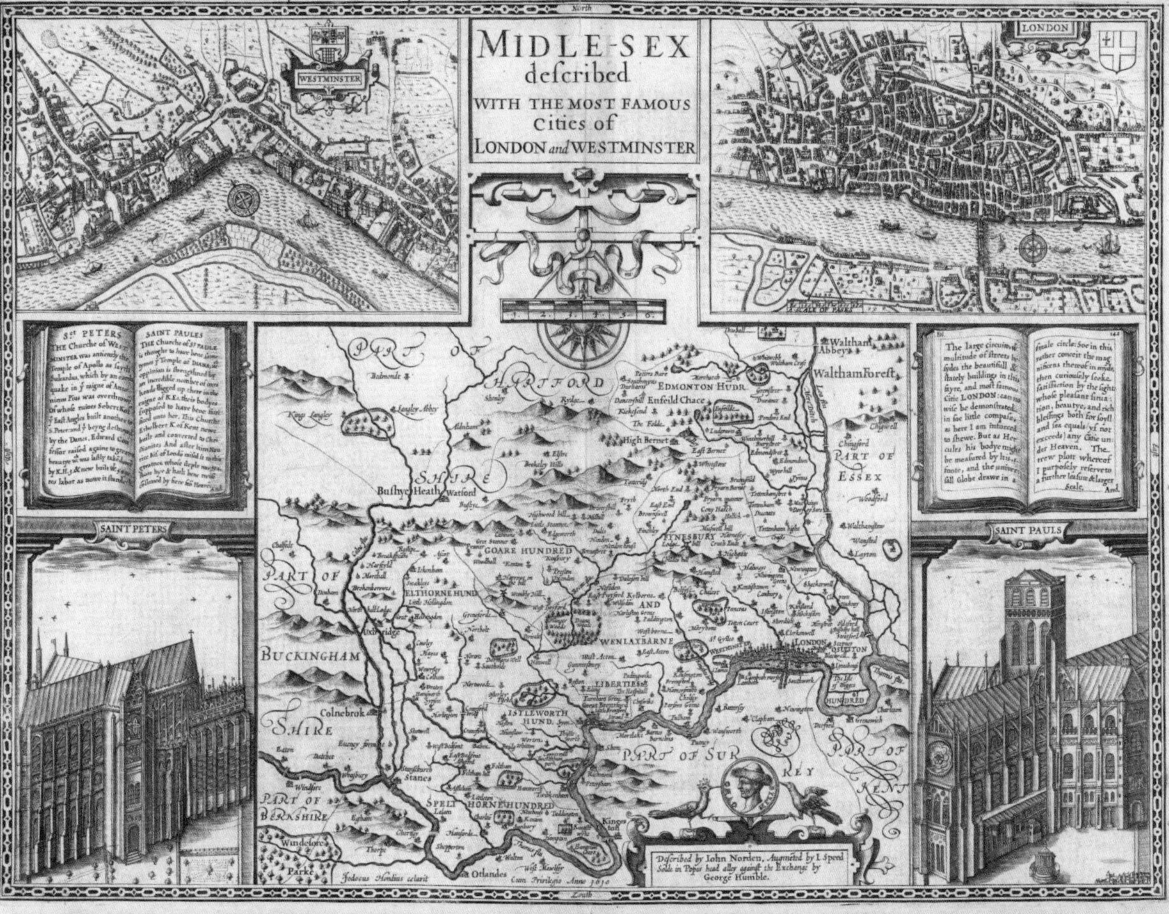
North
MIDLE-SEX
described
WITH THE MOST FAMOUS
Cities of
LONDON and WESTMINSTER
WESTMINSTER
LONDON
The large circuit of multitude of streets besydes the beautifull & stately buildings in this shyre, and most famous Citie LONDON: can no wise be demonstrated in soe little compasse, as here I am inforced to shewe. But as Hercules his bodye might be measured by his foote, and the universall Globe drawe in a
smale circle: Soe in this rather conceit the magnificens thereof in myde, then curiously seeke satisfaction by the sight whose pleasant situation, beautye, and rich blessings both for soyll and sea equals (yf not exceeds) any Citie under Heaven. The trew plott whereof I purposely reserve to a further leasure & larger Scale.
SAINT PETERS
SAINT PAULS
PART OF HARTFORD SHIRE
PART OF BUCKINGHAM SHIRE
PART OF BERKSHIRE
PART OF ESSEX
PART OF SURREY
PART OF KENT
EDMONTON HUDR
Enfeild Chace
High Bernet
Walthan Abbey
Waltham Forest
Bushye Heath
Watford
GOARE HUNDRED
ELTHORNE HUND
FYNESBURY
WENLAXBARNE
LIBERTIES
ISTLEWORTH HUND.
SPELT HORNE HUNDRED
LONDON
Colnebrok
Stanes
Uxbridge
Windesore
Parke
Otlandes
Kingston
Jodocus Hondius caelavit
Cum Privilegio Anno 1610
Described by Iohn Norden, Augmẽted by I. Speed
Solde in Popes head alley against the Exchange by
George Humble.
South

Elizabethan London depicted

8 **VERTUE, George, [after] Ralph AGAS**

Civitas Londinum Ano Dni Circiter MDLX. Londinum Antiqua. This Plan shews the ancient extent of the famous Cities of London and Westminster as it was near the begining of the Reign of Queen Elisabeth these Plates for their great scarcity are re-ingraved to Oblidge the Curious...

Publication
[London], George Vertue, 1737.

Description
Engraved plan on eight sheets.

Dimensions
(if joined) 2040 by 760mm
(80.25 by 30 inches).

Scale
28 inches to 1 statute mile.

References
Howgego 8, pp. 10–12.

A large, highly detailed re-creation of Elizabethan London. As is explained here, although this work was actually published in the eighteenth century, it is still one of only three known printed depictions of London from the sixteenth century. It captures the City as it was developing beyond its original walls, but with farms and pastures still much in evidence nearby. Deer can still be seen in St. James's Park. The major roads of entry to the city are shown and named, many of which are now well-known thoroughfares within the city. Many streets shown in the plan also bear names familiar to us. Much of the City of London's wall is still intact, and the Tower complex is well detailed. At the very lower left is Westminster Abbey, and just to the north is Scotland Yard, which contains a curious, smoking, dome structure, perhaps for the heating of tar for caulking ships. St. Paul's Cathedral is shown without a steeple, which fell in 1561, offering some corroboration to Vertue's 1560 date as the time depicted by the plan. Bear and bull baiting rings can be seen in the plan on the south bank of the Thames across from the City.

Although produced in the eighteenth century, this work was based on an extremely rare, sixteenth century plan, known now, according to Howgego, in just three examples. In fact, the original, attributed to Ralph Agas by Vertue on this plan, is referred to as being of "great Scarcity" in the title of this map. There is some doubt as to whether Agas was the maker of the original plan, as Vertue is the sole source for that attribution. Howgego states that the eight sheets that make up this plan were engraved in pewter, a metal very rarely used in the production of maps. George Vertue, who identified himself as an antiquarian on this work, first exhibited this plan at the Society of Antiquaries in London on March 21, 1737.

Finsburie Co
dogge hous
S Giles
More gate
All Holows in the Wall
Siluer str
Adle str
S Alban
Aldermond Bury
Colman streate
S Augustin
Broade streat
Lothbur
S Anthony
Chepesyde
Foster la
Mylke streate
S Laurence la
Ironmonger la
Olde Iurie
stokes
Buclersbury
S Pancras lane
Soper la
Wool Chur
Watlinge streate
Maidenhead lane
Basing lane
Thomas la
Turnebac la
budge row
Walbroke
lombard str
london ston
Canwicke St
S Thomas Apostle
Bread str
Garlik hyll
Whytyngton Colledge
Dowe gate
Lambert hyll
Broken Wharfe
Quenehyue
The Crane
Stylliarde
T H A M E S
Bull Bayting
Bear Bayting
Winchester Pl
S mary Ouer
see Letter . A . B . C . D . E .

The Newest & Exactest
MAPP of the most Famous Citties LONDON
and WESTMINSTER with their Suburbs; and the
manner of their Streets: With the Names of the Chiefest of them
Written at Length and Numbers set in the rest in sted of Names
The which Names are at Length in the Table With Numbers
how to finde them Readily So that it is a ready Helpe or
Guide to direct Countrey-men and Strangers to finde the
nearest way from one place to another. by T: Porter.
Printed & sould by Robt: Walton at the Rose & Crowne
at the West end of St. Paule's.
Pest house
Water house
old Street
St John street
Charter house
Barbican
St Barrs
West Smith feild
Cripple gate
Little Britain
Holborne
Grais Inn Lane
Grais Inn
Kings gate
Varnichhouse
St Gyles
Long acre
Queene street
Clare Street
Pietze
Lincolnes Inn feilds
Butchers
Strand
St Clements
Fleet street
Temple Barr
Condit
Lud gate
Newgate
Market
Cheapside
Paule church yard
Pecadilly Hall
Military garden
Gaming house
Pell Mell
St Jameses
St Jameses parke
White hall
Northampton house
Durham house
Salisbury house
Savoy
Somerset house
Strand bridge
Arundell house
Milford Lane
Essex house
THE
Temple stayres
White fryers
Bridewell
Black fryers
Baynards castle
Pauls wharffe
Broken wharffe
Queen hith
RIVER
Thames st.
The Banke side
Theeving lane
Tuttle Streete
old Palace yard
Lambeth Marsh
Lambeth house

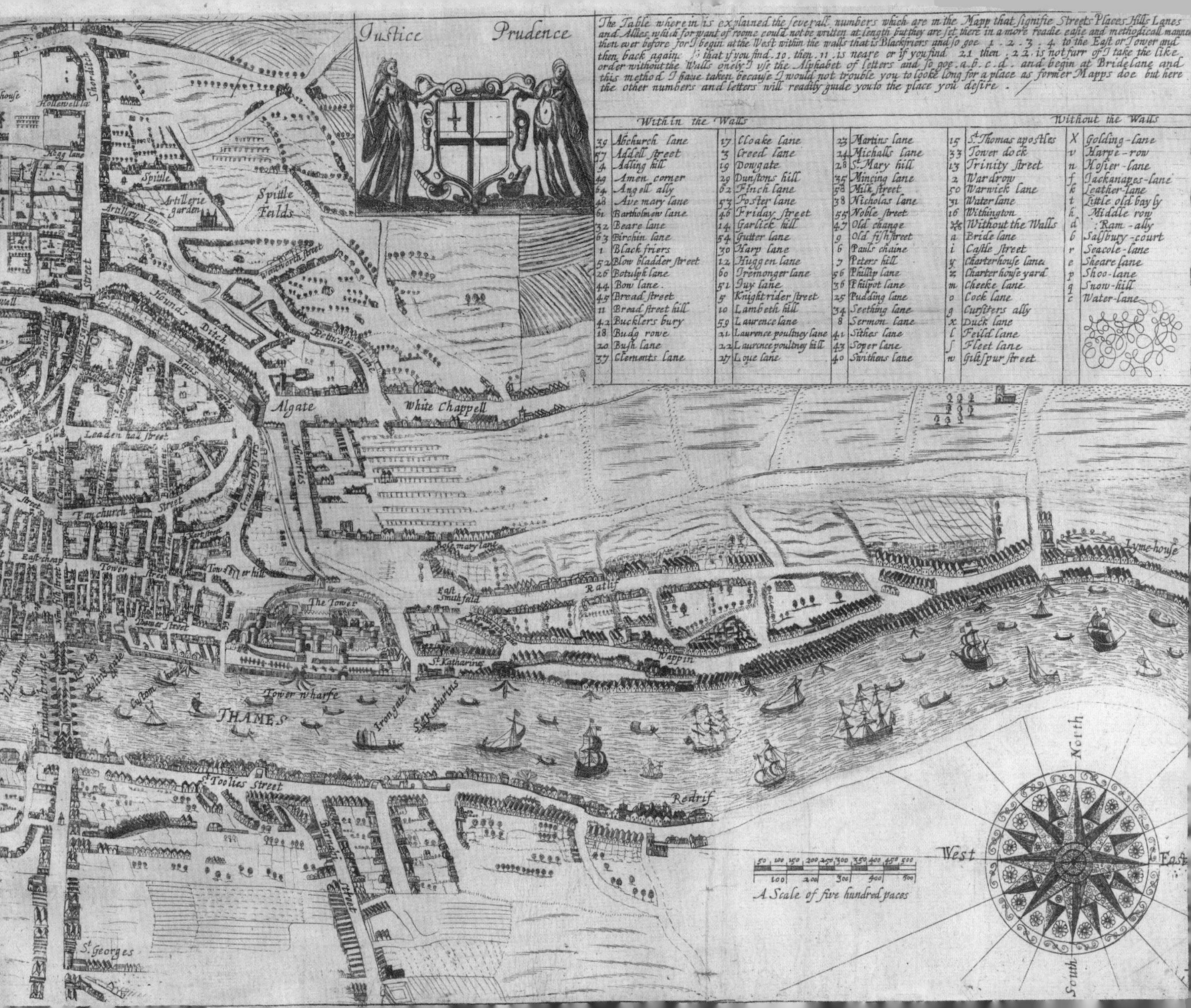
Justice
Prudence
The Table wherein is explained the severall numbers which are in the Mapp that signifie Streets Places Hills Lanes and Allies which for want of roome could not be written at length but they are set there in a more readie easie and methodicall manner then ever before for I begin at the West within the walls that is Blackfriers and so goe 1 . 2 . 3 . 4 to the East or Tower and then back againe so that if you find . 10 . then . 11 . is neare or if you find 21 then . 22 . is not farr of I take the like order without the Walls onely I use the Alphabet of letters and so goe . a . b . c . d . and begin at Bridelane and this method I have taken because I would not trouble you to looke long for a place as former Mapps doe but here the other numbers and letters will readily guide you to the place you desire .
Within the Walls
39 Abchurch lane
57 Addell street
4 Adling hill
49 Amen corner
64 Angell ally
48 Ave mary lane
61 Bartholmew lane
32 Beare lane
63 Birchin lane
1 Black friers
52 Blow bladder street
26 Botulph lane
44 Bow lane.
45 Bread street
11 Bread street hill
42 Bucklers bury
18 Budg rowe
20 Bush lane
37 Clements lane
17 Cloake lane
3 Creed lane
19 Dowgate
29 Dunstons hill
62 Finch lane
53 Foster lane
46 Friday street
14 Garlick hill
54 Gutter lane
30 Harp lane
12 Huggen lane
60 Iremonger lane
51 Iuy lane
5 Knightrider street
10 Lambeth hill
59 Laurence lane
21 Laurence poultney lane
22 Laurence poultney hill
27 Loue lane
23 Martins lane
24 Michalls lane
28 St. Mary hill
35 Mincing lane
58 Milk street
38 Nicholas lane
55 Noble street
47 Old change
9 Old fish street
6 Pauls chaine
7 Peters hill
56 Phillip lane
36 Philpot lane
25 Pudding lane
34 Seething lane
8 Sermon lane
41 Sithes lane
43 Soper lane
40 Swithens lane
15 St. Thomas apostles
33 Tower dock
13 Trinity street
2 Wardrow
50 Warwick lane
31 Water lane
16 Withington
Without the Walls
a Bride lane
i Castle street
y Charterhouse lane
z Charterhouse yard
m Cheeke lane
o Cock lane
g Cursiters ally
x Duck lane
l Feild lane
s Fleet lane
w Giltspur street
X Golding-lane
v Harpe-row
n Hosier-lane
f Jackanapes-lane
k Leather-lane
t Little old bayly
h Middle row
d Ram-ally
b Salsbury-court
r Seacole-lane
e Sheare lane
p Shoo-lane
q Snow-hill
c Water-lane
Shoreditch
Street
Hollewell la.
Hogg lane
Spittle
Artillerie garden
Artillery lane
Spittle Feilds
Wentworth street
Hounds Ditch
Petticote Lane
Algate
White Chappell
Leaden hal street
Minories
Fanchurch Street
East-cheap
Tower street
Tower hill
Rosemary lane
East Smithfeld
The Tower
Ratlif
Lyme-house
Wappin
St. Katharins
Tower wharfe
Custom house
Belins gate
Old swan
London bridge
THAMES
Irongate
St. Katharins
St. Toolies street
Redrif
Barnabi street
St. Georges
North
West
East
South
A Scale of five hundred paces

Thomas Porter's exceptionally rare plan of London

9 **PORTER, T[homas]**

The Newest and Exactess Mapp of the most Famous Citties London and Westminster with their Suburbs

Publication
London, Printed & sould by Robt. Walton at the Rose & Crowne at the West end of St. Paules, [c.1655].

Description
Engraved plan printed on two sheets joined, trimmed to neatline.

Dimensions
290 by 760mm (11.5 by 30 inches).

Scale
Scale: approx. 6 inches to 1 statute mile.

References
Howgego 11 (1).

The map is probably based upon a copy of the 'Ryther' map, now in the Bodleian Library, which has an extension – in pen and ink – westwards to Pall Mall and Lambeth House. The omission of the Stuart's royal coat-of-arms would suggest that the map was printed at some point in the 1650s. Howgego suggests a date of 1655, as the imprint upon the map is the same as present in Thomas Porter's 'New Book of Maps…' published in 1655. To the upper left is the long title, set within an elaborate title cartouche. In the title, the wish is expressed that the map "is a ready helpe or Guide to direct Countrey-men and Strangers to finde the nearest way from one place to another". To the right are the St George's Cross, the Harp of Ireland, and the Cross of St. Andrew. Further right, the City's arms are flanked by personifications of Juctice and Prudence, and a list of important "Streets, Places, Hills, Lanes, and Allies (sic) which for want of room could not be written". To the lower right is an elaborate compass rose and scale bar. The plan itself is finely engraved and shows Westminster and Lambeth House (Lambeth Palace) in the west to Lyme-House (sic) in the east, Pall Mall is written "Pell Mell", and the grand houses and wharves along the Thames are named. Upon the Thames, ferrymen, barges and sailing vessels are shown.

Only two other examples of the map are known: the one in the British Library's Crace Collection (Crace 34), differs from the present copy by the naming of "Bowe Street" in Westminster. Upon the present plan the street is named "Theeving Lane", which was the street's earlier name; therefore the present copy predates the Crace copy. The other example, housed at the Society of Antiquaries, is an even later state, and bears the royal coat-of-arms. It was later reproduced in facsimile by the London Topographical Society in 1898 (see item 10).

Item 9 (detail)

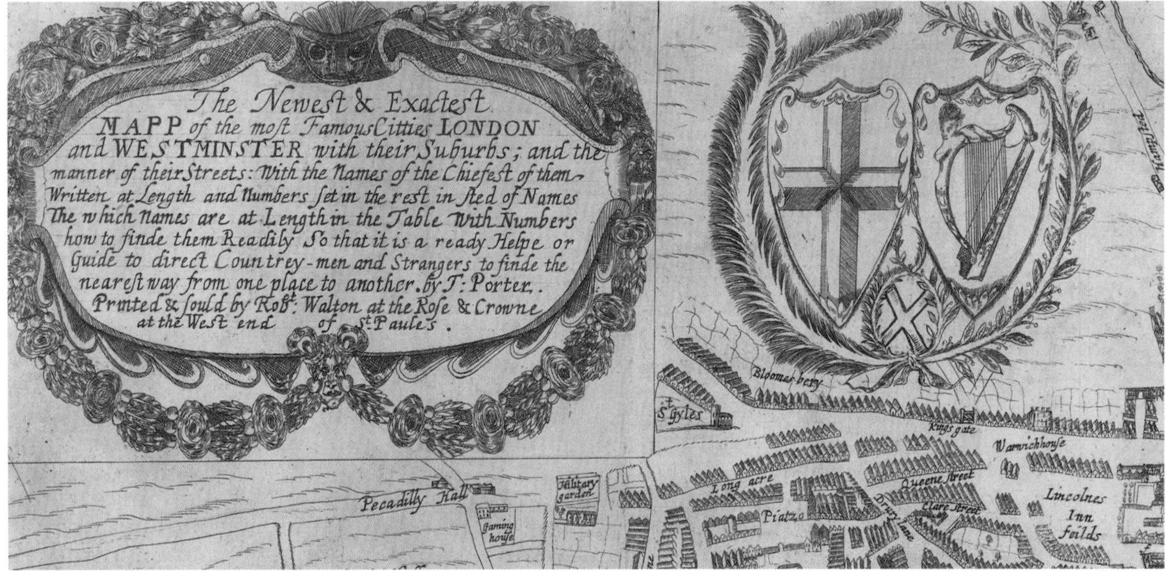

10 **PORTER, T[homas]**

The Newest & Exactest Mapp of the most Famous Citties London and Westminster.

Publication
London Topographical Society, 1898.

Description
Engraved plan.

Dimensions
300 by 750mm. (11.75 by 29.5 inches).

The original for this facsimile is in the Society of Antiquaries, and represents the third state described on the previous page (item 9).

Item 10 (detail)

The Great Fire

11 **DOORNICK, Marcus Willemz**

Platt Grondt der Stadt London. Amsterdam, Marcus Willemsz Doornick Boekverkoper op den Vygendam, 1666.

Description
Broadsheet engraved plan, letterpress text in Dutch, French, and English below, minor loss to imprint.

Dimensions
530 by 555mm (20.75 by 21.75 inches).

Scale
4¾ inches to 1 statute mile.

References
Howgego 17 (1).

The map was published in Amsterdam in the year of the Great Fire of London, and shows the devastation it wrought. Below the map is an accompanying text in Dutch, French, and English. The text gives a detailed account of the blaze, from its inception in a "Bakers howse in Pudding Lane" (marked with an 'A' on the plan), on Sunday morning of the 12th of September, to when it was "wholy extinguished and quenched on Thursday by reason of diverse Howses were blown up by powder, so that the flame was caused to desist". The fire had led to the complete destruction of all houses from the Tower of London to Temple. The text also lists all the streets that were affected by the fire.

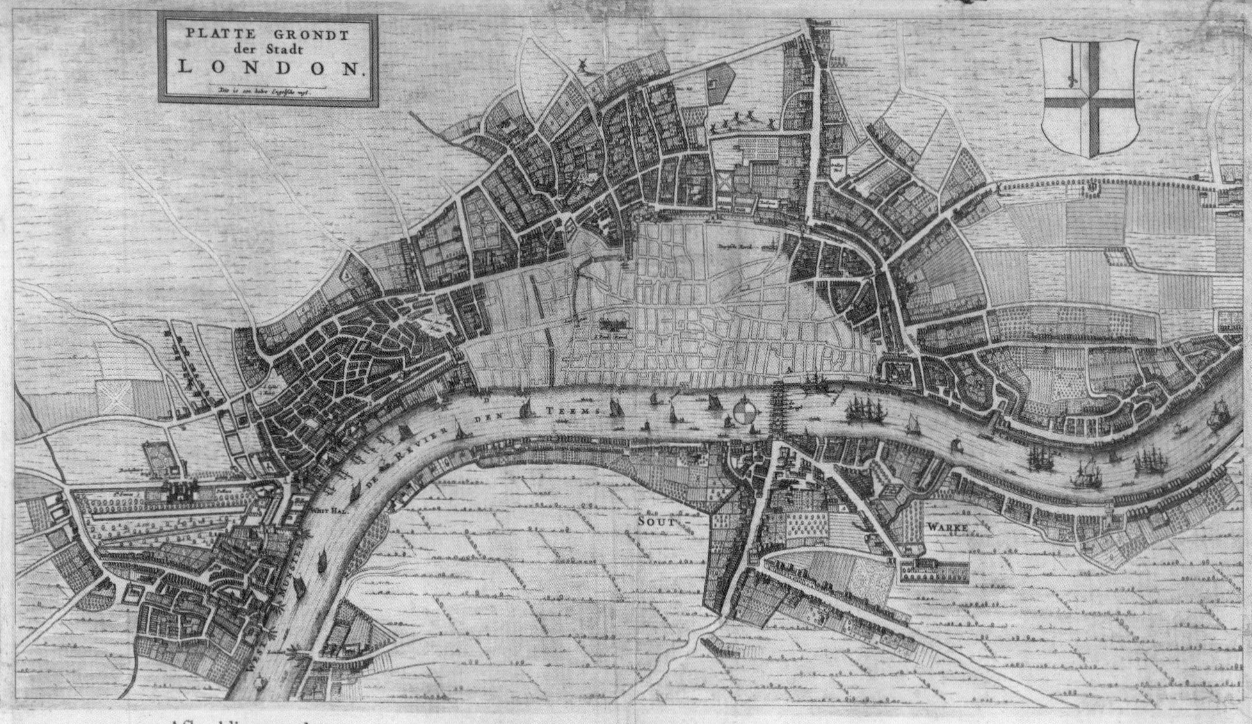

Afbeelding van de
STADT LONDON.
Aenwijzende hoe verre de zelve verbrandt is, en wat plaetzen noch overgebleven zijn.

Hoedanigh het meeste gedeelte der Stadt London, op den 12/13/14/15 en 16 Sept. des jaers 1666. door 't vuur is vernielt geworden/ en hoe verschrickelijk 't selve verwoest is/ wort in deze Afbeelding vertoont en beschreven; gelijk de zelve mede aenwijst de plaetsen die verbrant/ en de gene die noch overgebleven zijn. De Brant is dan eerst ontsteken den 12 Sept. des Sondaghs 's morgens tusschen 2 en 3 uren/ in een Bakkers Huys in de Pudding lane, dicht by de groote Brug, en wiert in 't begin weynigh geacht/ maer verspreyde sich door een stijve Noord-ooste windt aengeblasen/ wel haest in d'omleggende Huysen/ die aldaer dicht aen malkander/ van hout/ plaester/ en leem getimmert zijn: als wanneer alle mogelijke middelen/ tot uytblussing/ te vergeefs wierden aengewent. Des Maendagh 's avonts was het vuur alderverschrikkelijkst/ want het meeste van d'oude Stadt, geleek als in een enkele vlam te staen. Des Woensdaghs scheen den brandt gestuyt te zijn/ maer brak wederom by den Tempel uyt/ en began aldaer op nieuw/ doch wiert Donderdags t'eenemael geblust en opgehouden/ om datmen eenige Huyzen met Buspoeyer dede opspringen; waer door de vlam niet verder vatten kon. Doen zagmen dat van den Touwr af/ tot Tempelbare toe/ zijnde wel een groote ure gaens van malkander gelegen/ langs de Reviere den Teems, niet een Huys was blijven staen: zoo datmen (na dat den rook opgehouden was) by Tempelbare op de straet staende/ den Touwr zonder verhindering beoogen kon. In al deze langte zietmen niet om laeg/ als assen en puyn hoopen; en om hoogh eenige weynige vervallen muuren/ uytgebrande Toorens en Kerken/ en een deel overgebleven/ zoo heele als halve schoorsteenen uytsteeken. In de breedte van de Stadt/ vernielde de vlam al het geene dat'er stont van de Rivier den Teems af/ Noordt-waert op/ tot aen de oude Muuren by More-fields, blijvende aen de Noordt-oost zijde/ een gedeelte behouden/ als Bisschop gate-straet, Leadenhal-straet, Aldgate-straet &c. En aen de Rivier den Teems, aen de Oost-zijde/ den Touwr, (daer de vlam al in de Poorten was/ doch gestut wiert) St. Catherins, en vorder. Aen de Zuydt-zijde vloog de vlam uyt de Oude/ in de Nieuwe Stadt/ tot by Pyecorner en Smith-field, door Holborne, vernielende de gantsche Fleed-straet, tot aen Tempelbare, alwaer het gestut wiert. Alsoo is meer dan twee darde deelen van d'oude Stad London verbrandt/ met 85 Kerken/ daer onder een van de grootste van gantsch Christenrijk/ genaemt S. Paulus: 30 Gilde-huysen, en van de 98 Parochien zijn maer 12 ongeschent gebleven; men gist ontrent dertigh duysent Huysen zijn vergaen/ daer wel by de vijftig duysent woningen of huys-gesinnen in waeren gerekent. En of schoon de afgebrande ruynte in dese Afbeelding niet heel groot schijnt te wesen/ soo is het nochtans die plaets/ dewelke de machtighste Kooplieden/ rijkste Winkels/ Pakhuysen/ Solders/ Kelders/ vol van Koopmanschappen/ in sich begreep. De schade van desen brandt wert wel op hondert millioen ponden gewaerdeert.

Aenwijsing van de Namen der verbrande Straten, Stegen, en Plaetsen in LONDON.

A. de Pudding'lane, daer den Brandt eerst begonnen is.
B. de Touwr-poort, daer den Brandt gestut is.
C. Tempelbare, daer den Brandt ook gestut is.
Al de Huyzen in de Straten en Stegen, langs de Rivier, aen 't water uytkomende, van den Touwr-poort B, af, tot aen Temple Staires D. toe, zijn verbrant, en voorts
1 Thames-straet.
2 Nieuwe Fischstraet.
3 Touwr-straet.
4 East-Cheape.
5 Cannon-straet.
6 Grace Church-straet.
7 Flemmens-straet.
8 Lombard-straet.
9 Walbroke.
10 Broad-straet.
11 Cornehil.
12 De Beurs.
13 Poultry.
14 Chepside.
15 Nieuw-gate-market.
16 Holborne.
17 Fleete-straet.
18 Groote Wood-straet.
19 Kleyne Wood-straet.
20 Guild-Hal, 't Stadthuys.
21 Cateaton-straet.
22 Milke-straet.
23 Old Jury.
24 Loth Burry.
25 Gracious-straet.
26 Thrry morton.
27 Thrioneedle-straet.
28 Bokleis burry.
29 Kolemans-straet.
30 Balinghale-straet.
31 Alderman Burry.
32 Noble-straet.
33 St. Martins.
34 S. Hambles.
35 Pater noster Rowe.
36 De groote S. Paulus Kerk, Straet, en Kerkhof.
37 Knightrider-straet.
38 Old Fis-straet.
39 S. Thomas Apostel-straet.
40 Franchauer-straet.
41 Budge Rowe.
42 Bread-straet.
43 Watling-straet.
44 Friday-straet.
45 Old Change.
46 Adling hil.
47 Wardrow.
48 Pauls chaine.
49 S. Peters hil.
50 Lambeth.
51 Breadstraet hil.
52 Garlik hil.
53 Whittington.
54 S. Marie hil.
55 De Laken-hal.
56 Petter lane, ofslough.
57 Water lane.
58 Schoe lane.
59 Seacole lane.
60 Fleet lane.
61 Ave Marie lane.
62 Creed lane.
63 Carter lane.
64 Foster lane.
65 Gutter lane.
66 S. Laurens lane.
67 Iremonger lane.
68 Bow lane.
69 Soper lane.
70 S. Sithes lane.
71 Huggen lane.
72 S. Swithens lane.
73 Abchurch lane.
74 S. Nicolas lane.
75 S. Clements lane.
76 Birchen lane.
77 S. Michaels lane.
78 S. Martins lane.
79 S. Laurence lane.
80 Bush lane.
81 Philpot lane.
82 Rood lane.
83 Louer lane.
84 Botulph lane.
85 Water lane.
86 Beare lane.
87 Touwrdoke.
88 Bisschopgate-straet, ten deele verb.
89 Lyme-straet, ten halven verbrant.
90 Fanchurch-straet, ten deele verbr.
91 Mincing lane, een gedeelte verbr.
92 Marke lane, de helft verbrant.
93 Seating lane, een kleyn deel verbr.
De Duytsche Kerk (in de Broadstraet) is blijven staen, schoon dat de Huysen rondtom de zelve verbrant zijn. Als ook den Bisschop van Londons Huys: en Grasham Colledge, in Bisschopgate straet, daer nu de Kooplieden haer Beurs-tijds houden.

Representation curieuse de l'embrasement de la
VILLE de LONDRES,
Avec une Demonstration exacte de ce qui en est demuré de reste.

Vous pouvés voir dans cête peinture, comme la plûpart de la Ville de Londres a esté reduite à neant, & presque tout à fait consumée par le feu dont elle fut attaquée le 12 du mois de Septembre de l'année 1666. Elle vous represente aussi au vray les places qui en ont esté brulées, & celles qui sont restées, & qui ont eschapé la fureur de cet incendie. Le feu commença le Dimanche 12 Septembre entre 2 & 3 heures du matin dans la maison d'un Boulanger, situé à Puddinglane, *prés du Pont, & estoit si petit dans son commencement qu'on n'en tenoit point de conte & qu'on ne s'empressoit par beaucoup à l'esteindre, soit que les voisins fussent encore comme ensevelis dans le plus fort de leur sommeil ou qu'ils n'en apprehendassent pas les suites; Mais le Vent de* Nordoost *qui souffloit avec impetuosité l'alluma d'une telle maniere, qu'avant qu'il fust presque jour il fut epandu par toutes les maisons voisines, qui n'estant basties que de plâtre & d'autres matieres fort combustibles, & outre cela fort serrées, furent bientôt reduites en cendres, quels remedes imaginables qu'on employât pour s'opposer au cours de cet embrasement, qui augmenta jusques au Lundi au soir, qui ayant atteint le plus haut de sa fureur, il fit voir toute la vieille Ville en un monceau de flammes ramassées tout en un, jusques au Mecredi qu'il commença à s'esteindre, & se ralumant en suite du costé du* Temple *y continua avec violence jusques au Jeudi matin, qu'on fit sauter un grand circuit de maisons avec de la poudre pour arrester les flammes & leur oster la matiere qui les pouvoit entretenir. Ce remede estoit fort violent, mais il fit plus d'effet que tous les autres dont on s'estoit deja servi, & arrestant le cours des flammes, esteignit entierement le feu; le vent chassant en suite la fumée, on s'apperceut qu'il n'estoit pas resté une seule maison le long de la Tamise, depuis la* Tour *jusques à* Tempelbar, *qui sont eloignées l'un de l'autre de plus d'une lieue de chemin, si bien que lors que la fumée fut entierement dissipée, de l'un de ces endroits on pouvoit facilement voir l'autre, sans rencontrer dans cête étendue aucune chose qui s'opposât à la veuë; tout y ayant esté reduit en cendres & en une confusion epouvantable de mazures, ne restant aux environs que quelques murailles à demi ruinées, des tours, des Eglises, & des cheminées à demi brulées, & fort peu de bastiments qui n'ayent ressenti les effets de cet incendie, qui dans la largeur de la Ville reduisit à neant tout ce qu'il y avoit depuis la Tamise, en tirant vers le Nord, jusques aux vieilles murailles tout contre le* More fields, *laissant seulement du costé du Nord-oost la* Biscopsgate-straet, *la* Leadenhal-straet, *la* Aldegate-straet, *& quelques autres ruës qui sont demeurées en partie; & le long de la Tamise du costé de l'Oost la Tour (où le feu s'estant pris aux portes en fut bientôt esteint, tant parce que c'est un bastiment tout de pierre, que parce qu'on s'y opposa des le commencement) le fauxbourg* Ste. Catherine. *Mais plus avant, du costé du Zuid les flammes ayant esté portées, de la vieille ville dans la nouvelle à travers* Holborne *jusques à* Pyecorner *& le* Smithfield *le feu y fit un desordre epouvantable, & mit sans dessus dessous toute la* Fleed-straet, *jusques à* Tempelbar, *où son cours fut entierement arresté. De sorte qu'il y a plus des deux tiers de la vieille Ville de brulée avec 85 Eglises, y compris celle de S. Paul, qui estoit estimée la plus grande de toute la Chrêtienté; 30 Maisons de Guildes ou Maitrises, & 20000 maisons particulieres où habitoient plus de 50000 familles. Enfin cet embrasement a esté si grand & si general que de 98 Parroices dont la Ville estoit composée, il n'y en a que 12 où le feu n'ait touché, & qui n'en ayent ressenti les tristes effets: & quoi que cet espace que vous voyez en feu dans cête peinture ne semble pas bien grand, c'est pourtant ce lieu qui contenoit autrefois les plus riches boutiques, & les Magazins le mieux pourveus de toute l'Angleterre, si bien que pour reparer cête perte on croit certainement que cent millions de livres sterlin auroient peine à y suffire.*

A. *Puddinglane*, où le feu a commencé.
B. La *Porte de la Tour*, où l'on a éteint le feu.
C. *Tempelbare*, où l'on a aussi éteint le feu.
Toutes les maisons des ruës & ruelles allant à la Tamise tout le long de la Riviere ont esté brulées depuis la *porte de la Tour* B. jusques à *Temple Staires* D.
La rue de l'Evêque ou la *Biscopsgate-straet*, a esté brulée en partie, comme aussi la *Lyme-straet*, à demy.
La *Fanchur-straet*.
Mincing-lane.
La *Marke-lane*, a esté à demi brulée & quantité d'autres.
l'Eglise Flamande située dans la *Broad-straet* est échapée de cet embrasement, quoi que les maisons des environs ayent esté entierement brulées.
L'Hostel de l'Evêque de Londres est aussi demuré en son entier, & dans la Biscopsgate-straet, le College de Grasham où les marchands tienent de present leur Bourse.
Vous pouvés voir sous la description Flamande & Angloise les lieux & les ruës qui ont esté entierement brulées jusques au nombre de 93.

Delineation of the
CITIE LONDON,
Shewing how far the said citie is burnt down, and what places doe yet remain standing.

In what sort the greatest part of the Citie London upon the 12, 13, 14, 15 and 16 Septemb. Anno 1666. was consumed by fire, and how dreadful the ruine thereof was, is set forth and exhibited in this delineation, pointing also at the several places burnt down, and those that yet remain standing. The fire beganne upon the 12 Sept. about 2 or 3 of the klok Sunday morning at a Bakers howse in Pudding lane neer the Bridge, beeing at first not much regarded, til it spread it self verie forcibllie, by reason of an high North-east wind, unto the adjacent howses which were thereabout built thick one upon the other of wood, plaister and lime; Then were all possible meanstaken in hand for the quenching thereof, but in vain. On Monday night the fire was most dreadfull, for the greatest part of the Old Citie did seem wholy to be on a flame. On Weddensday the fire seemd to be stopt, but it brake out again by Temple-bar, beginning there afresh, but was wholy extinguished and quenched on Thursday by reason diverse Howses were blown up by powder, so that the flame was caused to desist. Then might be seen, that from the Tower unto the Temple, beeing aboue an houres going along the Thames, not one howse standing: Insomuch that the smoke ceasing, one might standing in the street by the Temple view the Tower without any let. In all this length of way downwards nothing is to be seen but ashes and heaps of rubbish, and upwards som few ruinous wals, consumed Towers and Churches with whole and half chimnies sticking out. In the bredth of the Citie the flame consumed all what stood from the River Thames Northwards unto the old Wals by Morefields, som part beeing preserved on the North-east side; with Bishopsgate street, Leaden-hal, Aldgate-street &c and at the River to the East-side the Tower (where the fire had entred the gates, but was stopt) S. Catharins and further. From the South-side the flame flew out of the Old Citie into the nieu unto Piecorner and Smithfield, through Holborne destroying al Fleet-street, unto Templebar where it was stopt. Thus were aboue two third parts of Old London burnt down, with 85 Churches (one whereof was the greatest of Christendom called S. Pawls) 30 Hals, and of 98 Parishes only 12 remaine entire. It is thought that about 20 thousand howses are demolished, in which were aboue fiftie thousand dwellings or families. And although the space burnt down seemeth not to be verie great in this Delineation, yet is it of that place, which had in it the mightyest Merchants, richest Ships, ware howses and Cellars ful of Marchandize. The dammage by this fire is judged to amount to an 100 millions of pounds.

Direction to the names of the Streetes and places burnt down in LONDON.

A. Puddinglane, where the fire first came.
B. The gates of the Tower, where the fire was stopt.
C. Templebarre where the fire was also stopt.
All the Howses in Streetes and Lanes along the River and comming out at the waterside from B. Tower-gate, to D. Temple Staires are wholie burnt down, and also,
1 Thames streete.
2 New Fish streete.
3 Tower streete.
4 East Cheape.
5 Cannon streete.
6 Grace Church streete.
7 Flemmens streete.
8 Lumberd streete.
9 Walbroke.
10 Broad streete.
11 Cornehil.
12 Old Exchange.
13 Poultrie.
14 Cheap side.
15 New gate Market.
16 Holborne.
17 Fleete streete.
18 Great Woodstreete.
19 Little Woodstreete.
20 Guild Hal.
21 Cateaton streete.
22 Milke streete.
23 Old Jurie.
24 Loth Burrie.
25 Gracious streete
26 Thrcy Morton.
27 Thrhineedle streete.
28 Baklers Burrie.
29 Colemans streete.
30 Basing Hale streete.
31 Alderman Burrie.
32 Noble streete.
33 St. Martins.
34 St. Hambles.
35 Pater noster Rowe.
36 S. Pawls Church, streete, yarde.
37 Knightrider streete.
38 Old Fish streete.
39 St. Thomas Apostle streete.
40 Franchaw streete.
41 Budge Rowe.
42 Bread streete.
43 Watling streete.
44 Friday streete.
45 Old Change.
46 Adling Hil.
47 Wardrow.
48 Pawls Chaine.
49 St. Peters Hil.
50 Lambeth.
51 Bread streete Hil.
52 Garlik Hil.
53 Whittington.
54 St. Marie Hil.
55 Cloth Hal.
56 Petter lane.
57 Water lane.
58 Shoe lane.
59 Seacole lane.
60 Fleet lane.
61 Ave Marie lane.
62 Creed lane.
63 Carter lane.
64 Foster lane.
65 Gutter lane.
66 St. Laurens lane.
67 Iremonger lane.
68 Bow lane.
69 Soper lane.
70 St. Sithes lane.
71 Huggen lane.
72 St. Swithens lane.
73 Abchurch lane.
74 St. Nicolas lane.
75 St. Clements lane.
76 Birchen lane.
77 St. Michaels lane.
78 St. Martins lane.
79 St. Laurence lane.
80 Bush lane.
81 Philpot lane.
82 Rood lane.
83 Lower lane.
84 Botulph lane.
85 Water lane.
86 Beare lane.
87 Tower dock, &c.
88 Bishopgate streete, partly burnt.
89 Lyme streete, halfburnt.
90 Fanchurch streete, halfburnt.
91 Mincing lane, partlyburnt.
92 Marke lane, partly burnt.
93 Southing lane, partly burnt.
The Dutch Church in Broadstreete, remaines standing (although the howses round about it were burnt down) and also the Bishop of Londons howse, and Grasham Colledge in Bishopgatestreet, where now merchants keep their Exchange meetings.

TOT AMSTERDAM,
By *Marcus Willemsz. Doorn* Boeckverkooper op den Vygendam. 1666.

London after the Great Fire

12 **LEAKE, John**

An Exact Surveigh of the Streets, Lanes and Churches Comprehending within the Ruins of the City of London. First Described in Six Plats, 10 Decemr. Ao. Domi 1666, by the Order & Directions of the Right Honourable the Lord Mayor, Aldermen, & Common Councell of the Said City, Iohn Leake, Iohn Iennings, Willm. Marr, Willm. Leybourne, Thomas Street, Richard Shortgrave, Svrveyors. & Reduced into One intire Plat.

Publication
[London, Engraved by George Vertue for the Society of Antiquaries], 1723.

Description
Engraved plan on two sheets, joined, hand-coloured in outline, a few old tears skilfully repaired.

Dimensions
500 by 1200mm (19.75 by 47.25 inches).

Scale
Approx. 17½ inches to 1 statute mile.

References
Howgego 21 (a).

Large and detailed plan of the City of London after the Great Fire.

Title on banderole across top. The map overlaps the banderole at top centre. Inset close to the top of the map are seven views of City buildings and streets. On the right is a map of London, Westminster and Southwark showing the burnt area of the City. On a banner below this the lines appearing on the map are explained. Scale-bar at bottom right. The map shows the extent of the Great Fire; the widening of some streets (e.g. Newgate, and Gracechurch Street to London Bridge); the creation of King Street and Queen Street and a Thames embankment; and the canalisation of the Fleet River.

Within days of the fire's extinguishing, new street layouts were being submitted to the king by architects including Christopher Wren, Robert Hooke, and the cartographer Richard Newcourt the elder. But before any plan could be implemented, accurate surveys were needed to chart the extent of the destruction. The king commissioned Wenceslaus Hollar and Francis Sandford to survey the City, and the corporation commissioned a team of surveyors, whose results were submitted on a plan drawn by John Leake. The plan was completed in March 1667 and engraved by Hollar, who also produced the seven inset views of important buildings and streets affected by the Fire: the Royal Exchange, Guild Hall, Cheapside & the Cross, St. Pauls, Temple Gate Fleet Street, and Baynards Castle.

The king put in place two principal measures to prevent future catastophic fires; namely, that all newly constructed buildings were to be built of brick and stone, and that streets were to be widened, so that if one side were on fire, the width of the street would prevent the other side from catching fire.

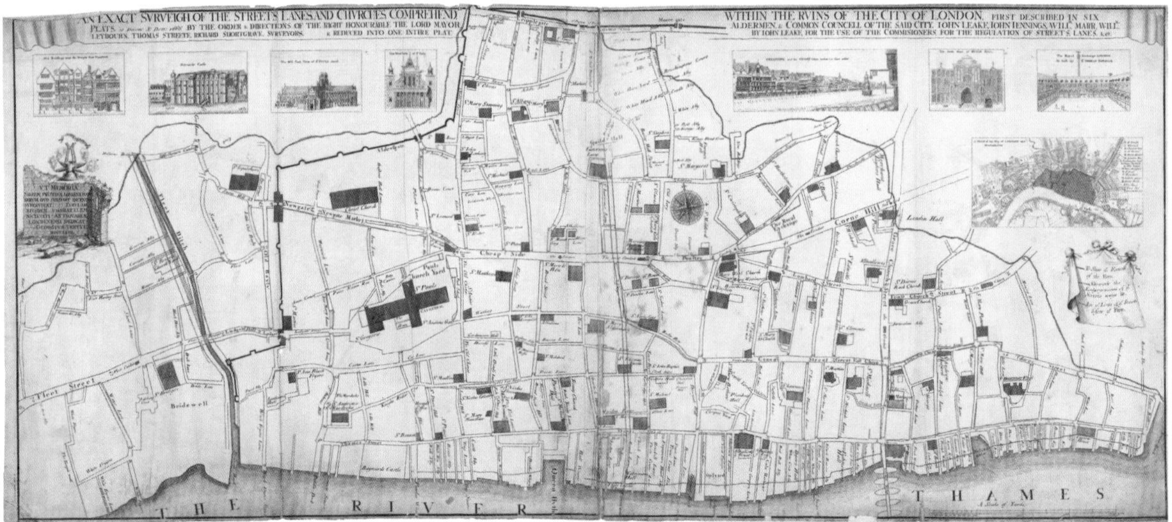

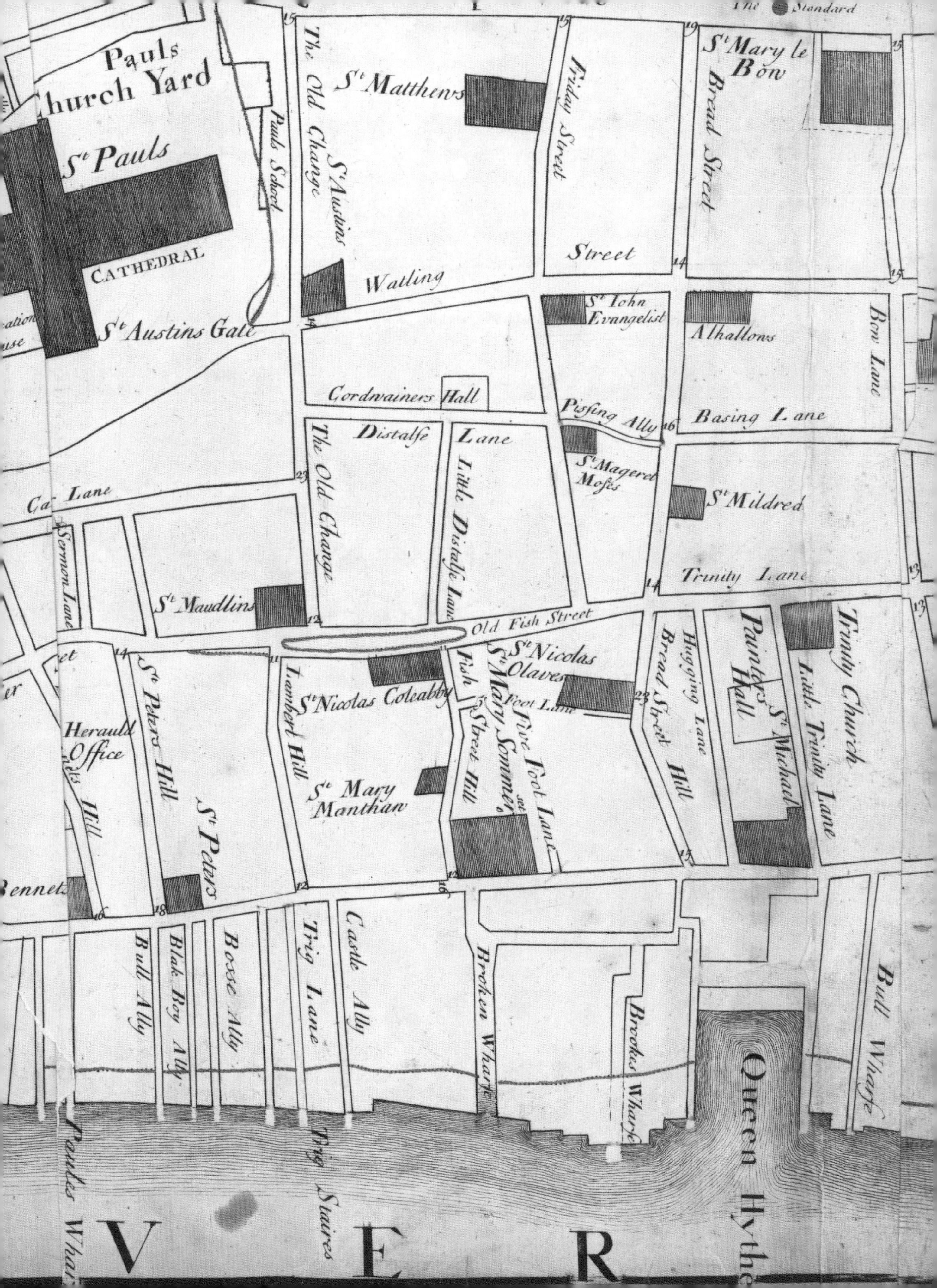

The Standard

Pauls
Church Yard
St Pauls
CATHEDRAL
St Austins Gate
Pauls School
The Old Change
St Austins
St Matthews
Friday Street
St Mary le Bow
Bread Street
Watling
Street
St John Evangelist
Alhallows
Bow Lane
Cordwainers Hall
Pissing Ally
Basing Lane
Distaffe
Lane
The Old Change
Little Distaffe Lane
St Magerett Moses
St Mildred
Car Lane
Sermon Lane
St Maudlins
Trinity Lane
Old Fish Street
St Nicolas Olaves
St Nicolas Coleabby
Fish
St Mary Sommerset
Five Foot Lane
Foot Lane
Fish Street Hill
Bread Street Hill
Hugging Lane
Painters Hall
St Michael
Trinity Church
Little Trinity Lane
Herauld Office
St Peters Hill
St Peters
Lambert Hill
St Mary Manthaw
Bennets Hill
Bennets
Bull Ally
Blak Boy Ally
Bosse Ally
Trig Lane
Castle Ally
Broken Wharfe
Brookes Wharfe
Queen Hythe
Bull Wharfe
Paules Wharf
Trig Staires
V E R

"The first accurate and detailed map of London"

13 **OGILBY, John and MORGAN, William**

[A New and Accurate Map of the City of London, distinct from Westminster and Southwark, Ichnographically Describin[g] all the Streets, Lanes, Alleys, Courts, Yards, Churches, Halls, Houses, &c. At the Scale of an Hundred Foot in an inch. Actually Survey'd and Delineated by John Ogilby Esq; and William Morgan, Gent. His Majesty's Cosmographers].

Publication
[London, Sold at the Author's House in White Fryers, 1676]

Description
13 sheets only (of 20, lacking the title, five northernmost, and the two lower south easternmost sheets), engraved map, each sheet approximately 400 by 515mm, sheets bear watermark of a coronet topped with a fleur-de-lys, some nicks to sheet edges, and some minor dust soiling.

Dimensions
1150 by 2500mm. (45.25 by 98.5 inches).

Scale
Approx. 52.75 inches to one statutory mile (purportedly 100 ft to the inch, or 1:1200).

References
Howgego 28; Barker and Jackson, pp. 38–41; Pennington 1007; cf. Dallaway, James, 'Inquiries into the origin and progress of the science of heraldry in England', Gloucester, 1793; Worms, Laurence and Baynton-Williams, 'British Map Engravers'.

In 1666, London was almost twice the size it had been at the moment of James I's accession, with much of the new development to the west of the City's boundaries. This left the City itself as a stinking, dark maze of streets, courtyards and alleys between the churches. It was this structure that the Great Fire cleared away in 1666. There was a requirement to rebuild the City in such a manner as to render another Great Fire impossible. Within a few days of the fire, three plans were presented to the king for the rebuilding of the city; those of Christopher Wren, John Evelyn and Robert Hooke. In the end, none of these plans was implemented, and the rebuilt city generally followed the pattern of the old one, much of which survives to the present day. When the Corporation of London's proposals came before Charles II in Council in 1667, they were returned with orders to plot the proposed streets on a map, so that the King might better be able to judge their breadth and give directions accordingly.

After the Great Fire, John Ogilby claimed, probably with some exaggeration, that he lost his entire stock of books valued at £3,000, as well as his shop and house, leaving him worth just £5. As he sought to restore his fortunes, Ogilby looked in new directions, and saw opportunity in the reconstruction of London's burnt-out centre. He secured appointment as a "sworn viewer", whose duty was to establish the property boundaries as they existed before the Fire. In this, Ogilby was assisted in the project by his step-grandson, William Morgan, and by a number of professional surveyors.

The result, whilst not printed until one month after Ogilby's death, was the present outstanding plan of London, on a scale of 100 feet to an inch.

"... nothing so precise was to appear again until the Ordnance Survey two centuries later" (Barker and Jackson)

"This is the first accurate and detailed map of London, with all the buildings represented in plan rather than as bird's eye views" (British Library Catalogue of the Crace Collection)

The plan was not Ogilby and Morgan's alone: Wenceslaus Hollar was responsible for a great deal of the engraving, and Gregory King, under direction of Ogilby, appears to have engraved large portions also.

The attribution to Hollar comes in part from a letter by George Vertue to Dr Ducarel, dated 1 September 1749: "Kind Sir, The same day I received your letter... The Plan of London and other workmen, went by the name of Ogilby's Plan of London; and after I published Hollar's works, I found it to be some part done by him; but nowhere on it his name or mark, therefore it has been omitted..."

Gregory King went to London in 1672 and met Hollar who recommended him to Ogilby "to manage his undertakings". King says he etched some plates for Ogilby and he claims to have been responsible for the scale and design of the 20-plate map of London which was "etched in copper by Mr Hollar" (King's autobiographical notes printed in Dallaway).

Cock Lane
Churchyard of St Bartholomew the less
Christ Hospitall
Old Church
Christ Church
Newgate
Newgate Street
Snow Hill
Green Arbour
Fleet Lane
Fleet
Old Baly
Sessions House yard
Warwick Court
Newgate Market
Pater Noster Row
Amen Corner
Stationers Hall
St Martin Ludgate Church & Yard
Ludgate Hill
Ludgate Street
Fleet Bridg
St Pauls Cathedral
St Pauls Church Yard
Blow Bladder Street
Watling Street
Holliday Court
Shoomaker Row
Carter Lane
the King's Wardrobe was here
St Andrew Wardrobe Church & Yard
The Kings Printinghouse
Knight Rider Street
Old Fish Street
Thames Street
St Peters Church
St Mary Somerset Church
Broken wharfe
Wood wharfe
Wharfe
Bride lane
New Canal
Well
River of Thames
A Scale of Feet
A 100 in An Inch

Robert Hooke was also heavily involved in the production, and makes numerous references to it in his diary. Indeed, read together, Hooke's diary entries give a vivid insight into the stages and timetable involved in preparing a large-scale survey and map:

14 AUG 1673	"At Armorers with Ogylby, at Garways [Garraway's coffee house] designd sheets for London"
16 AUG 1673	"At Mr Holler [Wenceslaus Hollar] concluded scale of 100 in an inch"
26 AUG 1673	"With Ogylby at Garways, told [Gregory] King his man of the new way of making batts to print with the rowl press"
22 SEPT 1673	"At Garways with Ogylby and Aubery [John Aubery]"
14 OCT 1673	"At Garways with Ogylby. Shewd him the way of letters for marking his map and also the way of shadowing"
8 DEC 1673	"With Ogylby at Dr Godders he brought me his 1st sheet of London from Hollar"
9 JAN 1674	"Mr Ogylby shewd me his second sheet and I wrote him a report to the court of aldermen"
15 JAN 1674	"With Mr Ogylby to Spanish coffee house… contrived pacing saddle with waywiser"
29 JAN 1674	"With Mr Ogylby at Guildhall about map of London"
19 MAR 1674	"At Spanish coffee house with Ogilby. Drew the uses of the London map"
26 JAN 1675	"with Ogilby at Joes. I was witness to a bond of his and Morgans to Brook, plaisterer, for £50 as he told me". Ogilby died 4 Sept 1676.

The 'Term Catalogues' for 22 November 1676 record that "The Survey of London nine feet long and six feet deep is to be sold by Will. Morgan his Majesty's Cosmographer at his house in White Friers. £1.5.0 in sheets. Also to be published at a less scale a map of London Westminster & Southwark actually surveyed; and as counterfeits have been made these will be known by the names of John Ogilby & William Morgan, with a scale to warrant them".

Robert Hooke evidently bought his copy at a discounted price and somewhat later: his diary on 14 Feb 1677 reads "Ogylbys map from Morgan. Gave his man 5sh".

Hooke was not the only diarist to make mention of the map: "Up, and to the office, where we sat all the morning, and my Lord Bruncker did show me Hollar's new print of the city, with a pretty representation of that part which is burnt, very fine indeed; and tells me that he was yesterday sworn the King's servant, and that the King hath commanded him to go on with his great map of the city, which he was upon before the city was burned, like Gombout of Paris, which I am glad of" (Samuel Pepys' Diary 22 November 1666)

In 1677 Morgan issued a printed pamphlet to accompany the map "London Survey'd: or, an explanation of the large map of London", of which there is an unique example in the British Library.

The map was re-engraved later, undated, and with the new title: "A Large and Accurate Map of the City of London Ichnographically Describing all the Streets, Lanes, Alleys, Courts, Yards, Churches, Halls and Houses, &c. Actually Surveyed and Delineated, by John Ogilby Eq; His Majesty's Cosmographer."

The Ogilby and Morgan plan represents very nearly the first linear ground plan or 'plot' of any British town. It was certainly the first large, multi-sheet plan of a British town to be so delineated. It depicts a London recovering from disaster, and is refreshingly frank in its assessment of the capital; prisons and working-class areas feature on the same scale as palaces and grand avenues (the same cannot be said of Morgan's later, 1682, plan of the city: a grand heavily decorated engraving devoid of poverty and places of incarceration). The part-built and planned edifices on the 1676 survey signify a level of hope and optimism, and the London that emerges from the ashes of the Great Fire, whilst familiar, has a new shape:

Fleet Street, the route from Temple Bar to St Paul's, divides the plan horizontally, and it is bisected north to south by the New Canal, with quays, 30ft wide, on either side of the now navigable Fleet River.

Although plans such as Evelyn's and Wren's were impracticable, the old layout of streets is much improved. Thames Street is straighter and wider, Ludgate from St Bride's to St Paul's is now 40ft across, even lanes and alleys are at least 14ft wide: enough for two drays to pass.

The Duke's Theatre, attributed to Wren, is shown on the waterfront.

The Tudor palace of Bridewell, now a prison, hospital and school, looms large over the New Canal, while further north is Fleet Prison where debtors – among them the playwright Wycherley – are held. Many sizeable gaps are yet to be filled and one, marked "The King's Wardrobe was here," is a reminder of the pre-fire storehouse for royal robes, now removed to The Savoy.

The most impressive building is the new Royal Exchange, larger than before and with more shops, completed in 1671, the same year in which the Guildhall is sufficiently repaired to host the Lord Mayor's Banquet again.

Of the 84 churches destroyed by the fire, 30 are now restored or well under construction, with Wren deciding which shall be saved, and which lost, and which parishes united.

New markets, such as Honey Lane, were designed in order to remove traders from Cheapside, and into large spaces to prevent them clogging up the roads: the Wool Church Market in the southeast for vegetables, and Newgate Market between Newgate and Paternoster Row, are two examples visble on the map.

HIGH HOLBORN
HOLBORN
Grays Inne
Charterhouse Yard
ST JOHN STREET
SMITH-FIELD
LINCOLNS INN
LINCOLNS INN FIELDS
LOWER LINCOLNS INN FIELDS
CHANCERY LANE
FLEET STREET
LUDGATE HILL
LUDGATE STREET
CHRIST CHURCH
Old Church
ST PAULS CATHEDRALL
PAULS CHURCH YARD
BRIDE WELL
FLEET
CHEAP
THE RIVER OF THAMES
A SCALE OF FEET

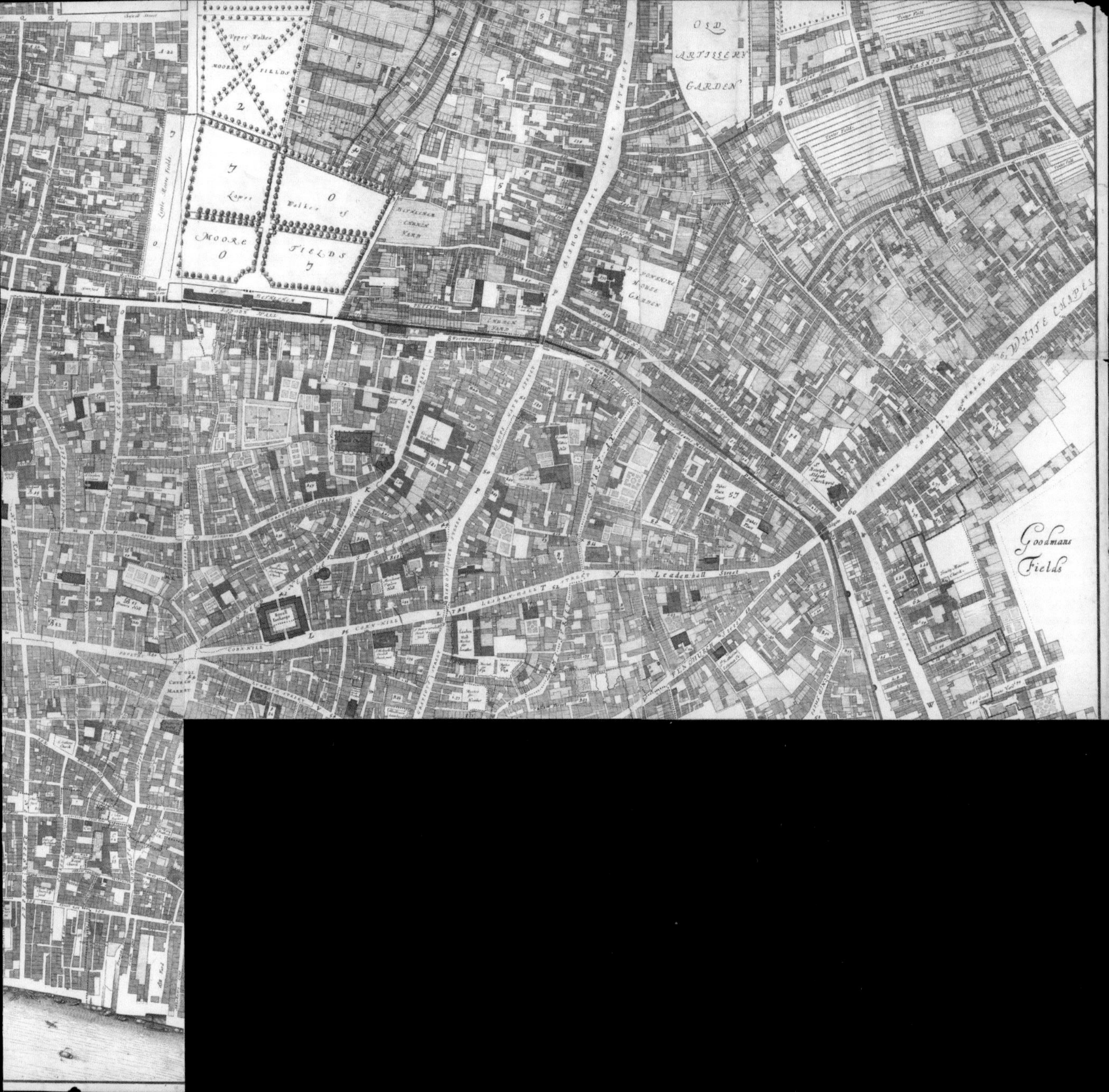

Upper Walkes of
MOORE FIELDS
OLD ARTILLERY GARDEN
Lower Walkes of
MOORE FIELDS
BETHLEHEM CHURCH YARD
Bishops Gate Street Without
DEVONSHIRE HOUSE GARDEN
WHITE CHAPEL
Leadenhall Street
CORN HILL
Royal Exchange
Goodmans Fields

England's glory

14 **WALTON, Robert**

England's Glory or The Glory of England, Being a New Mapp of the Citty of London Shewing the remarkable streets Lanes Alleyes Churches Halls Courts and other places as they are now rebuilt the which will therefore be a guide to Strangers and such as are not well acquainted herein to direct them from place to place. Divers faults yt are in ye former are in this amended. Allso the severall figures yt stand up and downe in ye Mapp are explained in ye 2 tables at ye upper corners hereof.

Publication
London, Printed and Sold by Robert Walton, at ye Globes Compasses, just at the west end of St. Paul's-Church, [1676].

Description
Hand-coloured engraved plan.

Dimensions
430 by 550mm (17 by 21.75 inches).

Scale
11 inches to 1 statute mile.

References
Howgego 30.

Detailed plan of the the City of London.

The plan extends from Fleet Street in the west to the Tower of London in the east and from Finsbury Fields in the north to Bankside in the south. To the upper left and right is a list of public buildings, gates, bridges, and streets. The plan shows in detail the rebuilt City of London which the Great Fire had destroyed much of in 1666. One can clearly see the heart of what is now the financial district including Cheapside, Cornhill, Lombard Street, and Leadenhall Street, together with the Monument, St. Paul's, and the Tower of London. In the title, Walton is keen to stress the usefulness of the plan being, "a guide to strangers, and such as are not well acquainted herein to direct them from place to place". He goes on to write: "Divers faults yt are in ye former are in this amended". To which plan Walton was alluding is unclear; some have suggested that it was his: "Newest and Exactess Mapp of the most Famous Citties London…", of 1655 (item 9) yet as this depicts a pre-Fire London it would seem unlikely. Most probably the plan is a revised state of an earlier, no longer extant, plan. This would also account for the rather haphazard placing of a large part of the explanatory text. The map can be reliably dated to 1676, as Walton advertised it for sale in the 'Term Catalogue' of May that year.

Glory of the Glory of England, Being A New Mapp of the City of LONDON
Showing the remarkable streets Lanes Alleyes Churches Halls Courts and other places
Printed and sold by Robert Walton at the Globe & Compasses just at the west end of St Pauls Church
Finsbury Feilds
Bun hill Feilds
Churchyard
The Artillery Ground
Charter house yard
Smith Feild
Holborn
Snow hill
Newgate St
Old Street
Goswell Street
London wall
Cheapside
Cornhill
Fleet Street
Tower Street
Tower hill
East Smithfeild
The Tower
The Temple
Spittle Feilds
Artillery Ground
Bishopsgate Street
White Chaple
The Minories
Fenchurch Street
King Street
Queen Street
Broad Street
Fish street
Thames Street
The River of Thames
London Bridge
Billingsgate
Dowgate
Temple Staires
Bridge

London as seen from overseas

15 **CORONELLI, Vincenzo Maria**

Londra Dedicata All'Illustrissimo S. Nicolao Cornaro Figliuolo dell'Excellentissimo Sigr. Procre. Francesco.

Publication
[Venice, 1689].

Description
Engraved plan.

Dimensions
360 by 485mm (14.25 by 19 inches).

Scale
4 inches to 1 statute mile.

References
Howgego 37.

The plan is flanked by the arms of the City of London to the left, and the the Royal Standard to the right. Below the plan is the title and dedication to Nicolao Cornaro, a leading member of the Venetian aristocracy. The title is surmounted by a coat-of-arms. Whose arms they are is uncertain, however, they are probably those of Cornaro. The only name to appear on the plan is that of the River Thames.

The plan was published in Coronelli's rare 'Citta, fortezze, isole, e porti principali dell'Europa in pianta…' in Venice 1689.

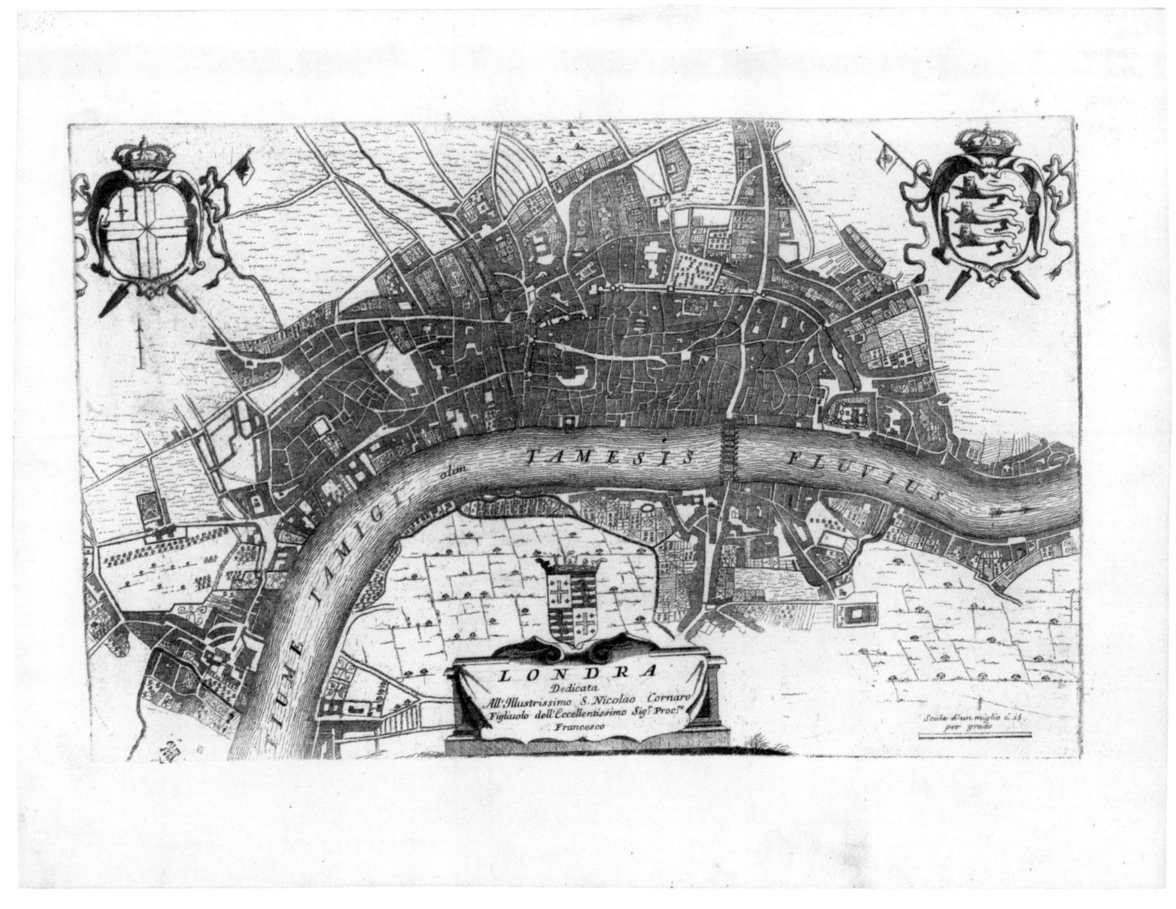

Rare plan depicting London during the reign of William and Mary

16 [Anonymous]

[London].

Publication
[?Amsterdam, c. 1690].

Description
Hand-coloured engraved plan, with 18 inset views of major public buildings, gates, and statues, a few nicks and tears skilfully repaired.

Dimensions
485 by 595mm (19 by 23.5 inches).

Scale
Approx. 6 inches to 1 statute mile.

References
Howgego 39.

This highly detailed plan extends from Westminster in the west to Wapping in the east and Clerkenwell in the north to St. George's Fields in the south. The most important public buildings are depicted in elevation, including the Tower of London, Westminster Abbey, and St Paul's. Several places are marked with letters and numbers. To the left of St George's Field is a reclining female, with a globe and pair of compasses; she sits upon the scale bar and is surrounded by putti holding surveying tools. To the upper left and right are the royal and City of London coat-of-arms respectively. Above and below the plan are eighteen views of public buildings, gates, and statues; they are from upper left to lower right: The Temple, Whitehall, Westminster Hall, The Royal Exchange, Clarendon House, Covent Garden, New Bedlam, Bridewell, Cripple-gate, Alders-gate, Newgate, Ludgate, Statue of Charles II at the Stocks Market, and Charles I at Charing Cross, The Tower of London, Moore-gate, Ald-gate, and Bishops-gate. Also along the upper border is a cameo of William and Mary.

Separately issued map of London and the surrounding counties

17 **LEA, Philip**

A Mapp Containing the Townes Villages Gentlemens Houses Roads Rivers Woods and other Remarks for 20 Miles Round London.

Publication
London, Sold by Phil Lea at ye Atlas & Hercules in Cheapside, [c.1690].

Description
Engraved map, original outline hand-colour, minor nicks and tears skilfully repaired.

Dimensions
Dimensions: 530 by 570mm (20.75 by 22.5 inches).

Scale
approx. ½ inch to 1 statute mile.

References
Howgego 41 (1).

The map gives details of market towns which – as the key tells us – are marked in a "Round Roman Hand", parish churches, villages and hamlets, gentlemen's houses, ordinary houses, roads and rivers. The map bears grid lines, with each square said to contain three statute miles.

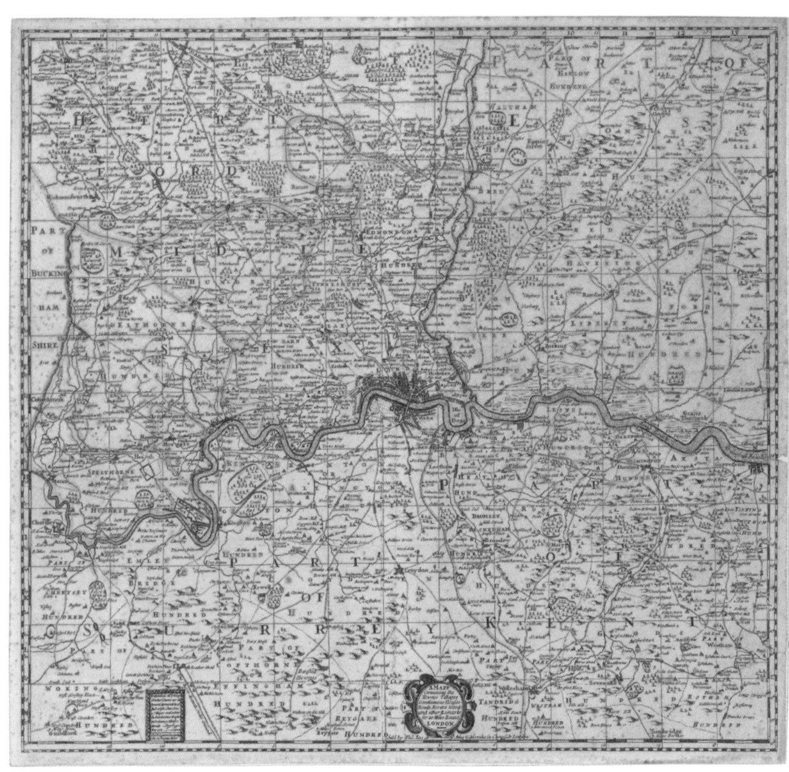

WALTHAM
Epping Forest
Epping Street
Waltham Abby
Chingford
Woodford
Barking
Ilford
West Ham
East Ham
Stratford
Bromley Hall
Dagenham Creek
Barking Creek
EDMONTON
HUNDRED
Enfeild
Enfield Chase
Barnet
Baram Wood
South Mimms
Stanmore
Edgworth
Finchley
Finchley Common
Hendon
Kingsbury
Harrow on y hill
FYNNESBURY
Highgate
Hornsey
Islington
Tottenham
Newington
Hackney
Pancras
Kensington
Hammersmith
Chelsey
Fulham
Putney
Wandsworth
Battersey
Clapham
Camberwell
Peckham
Dulwich
Lambeth
Newington
LONDON
SOUTHWARK
Westmin
Stepney
Popler
Isle of Dogs
Deptford
Greenwich
Woolwich
Charlton
Eltham
Leusham
Shooters Hill
New Branford
Richmond
Kingston
New Park
Mortlack
Wimbledon
Streham
Mitcham
Croydon
Carshalton
Banstead
Beckenham
Bromley
Chislehurst
St Mary Cray
Pauls Cray
Kestion
Orpington
Farnborough
Cudham
Erwell
Cheame
Ewell
Sutton
Woodcote
Purley
Sanderstead
Chelsham
Warlingham
BRIXTON
ALLINGTON
HUNDRED
PART OF
SURREY
KENT
MIDDLESEX
The Road

Morden and Lea's large and attractive plan of London and the surrounding country

18 **MORDEN, Robert and LEA, Philip**

This Actuall Survey of London, Westminster & Southwark is Humbly Dedicated to Ye Ld Mayor & Court of Aldermen. By Ro. Morden, Phil. Lea, Chr. Browne.

Publication
London, Sold by Philip Lea at the Atlas & Hercules in Cheapside and by Christopher Browne at the Globe the west end of St. Paul's, 1700.

Description
Engraved plan on two sheets joined, extending north to south from St Pancras to Southwark, and from west to east from Arlington House (Buckingham Palace) to Stepney, a few old folds reinforced, a few tears skilfully repaired, and some minor dampstaining to banner.

Dimensions
655 by 1030mm (25.75 by 40.5 inches).

Scale
7½ inches to 1 statute mile.

References
Howgego 42 (2).

The most important public buildings are shown in elevation. Amongst them are some notable lost palaces, including Arlington House, on the present day site of Buckingham Palace. The house, built in 1675, was the residence of Henry Bennet, 1st Earl of Arlington. The freehold for the property would be sold to the Duke of Buckingham in 1702. Another palace of note is Montague House, home of the 1st Duke of Montague. It was situated on Great Russell Street, which at the time backed on to fields, and in 1759 was sold to the Trustees of the British Museum. The house would be demolished in the 1840s to make way for the modern day building.
A great deal of the outlying area of the map is a patchwork of fields dotted with hamlets, such Marylebone, St Pancras, and Cambridge Heath.
Below the plan are tables giving names of the public offices, wards, and parishes; which before the Fire numbered 97 but only 62 after. There is also information on the halls and companies, markets, inns of court, prisons, palaces, and public buildings. The Thames is depicted teeming with sailing vessels, barges and watermen.

This Actuall Survey of LONDON, WESTMINSTER & SOUTHWARK is humbly Dedicated to ye Ld. Mayor & Court of Aldermen by
Rt. Morden
Phil. Lea
Chr. Browne
THE RIVER OF THAMES
HALLS & COMPANYES
MARKETS
PRISONS
PALLACES
HOSPITALLS

Cavendish Square under development…

19 **COVENS, Johannes and Cornelius MORTIER**

This Actuale Survey of London and Westminster & Soutwark [sic]

Publication
Amsterdam, Iean Covens et Corneille Mortiers. Libraires et Marchands des Cartes, [c. 1725].

Description
Engraved plan on two sheets joined, a few old folds reinforced.

Dimensions
655 by 1035mm (25.75 by 40.75 inches).

Scale
7½ inches to 1 statute mile.

References
Howgego 42 (4).

This map is a reissue of a 1690 plan published by Morden and Lea. The dedication and title are on a banderole across the top of the plan, with the City of London arms and compass rose at top center. Below the plan is a key to public offices, wards, parishes within the city walls, which numbered 97 before the Great Fire and 62 after, halls, companies, markets, prisons, and hospitals. Prominent buildings are shown in elevation. Cavendish Square is shown as under development.

...Westminster Bridge depicted

20 **COVENS, Johannes and Cornelius MORTIER**

This Actual Survey of London, Westminster & Soutwark [sic]

Publication
Amsterdam, Chez Iean Covens et Corneille Mortier, Libraires et Marchands des Cartes, [c.1755].

Description
Engraved plan, on two sheets joined, tear to join at foot not affecting image.

Dimensions
650 by 1070mm (25.5 by 42.25 inches).

Scale
7½ inches to 1 statute mile.

References
Howgego 42 between (4) and (5).

The plan is an intermediate state between Howgego 42 states (4) and (5): the spelling of 'Actual' is the same as in state (4); Westminster Bridge is depicted, which was completed in 1750, but not Blackfriars Bridge (as is shown on state (5)), which was begun in 1760.

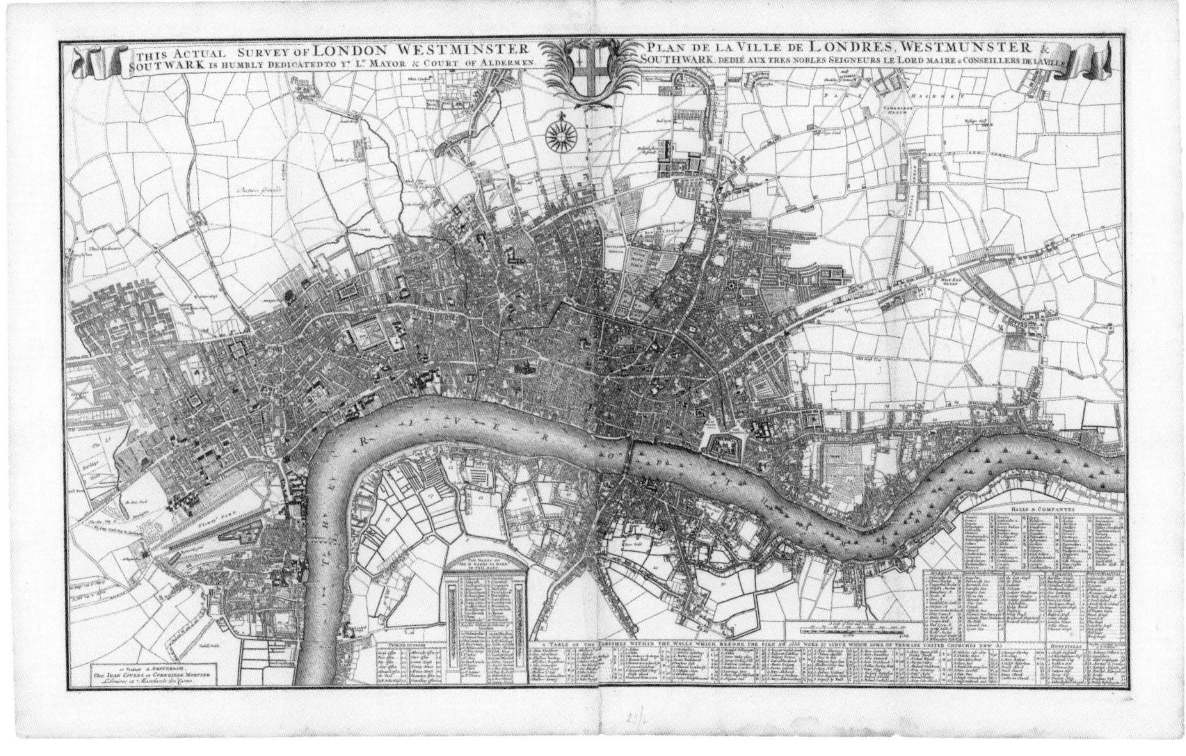

"nearly as big as Paris"

21 **FER, Nicolas de**

Plan des villes de londres et de westminster et de leurs faubourgs avec le bourg de southwark. avec Priv. du Roy 1700.

Publication
[Paris, 1700].

Description
Hand-coloured engraved plan.

Dimensions
290 by 330mm (11.5 by 13 inches).

Scale
Approx. 3 inches to 1 statute mile.

References
Howgego 44 (1).

French cartographer, geographer, engraver and publisher Nicholas De Fer (1646–1720) took over the business begun by his father Antoine De Fer. Nicholas was a prolific mapmaker and publisher of over 600 sheet maps, wall maps and atlases. His maps were prized for their decorative qualities rather than the accuracy of their geography. Nonetheless, his reputation grew, culminating in his appointment as Geographer to the King. Among his works are several atlases: 'France Triomphante' in 1693, 'Forces de L'Europe' 1696, 'Atlas Curieux' 1705 and 'Atlas Royal'.

"At the opening of the eighteenth century … French maps of London were for the most part small and unimportant, being derived from native originals, but it is interesting to find a London map of 1700 published by a Parisian, Nicolas de Fer, being used as a means of showing his pride in his native city. London he says, is "nearly as big as Paris". (Howgego, p. 19)

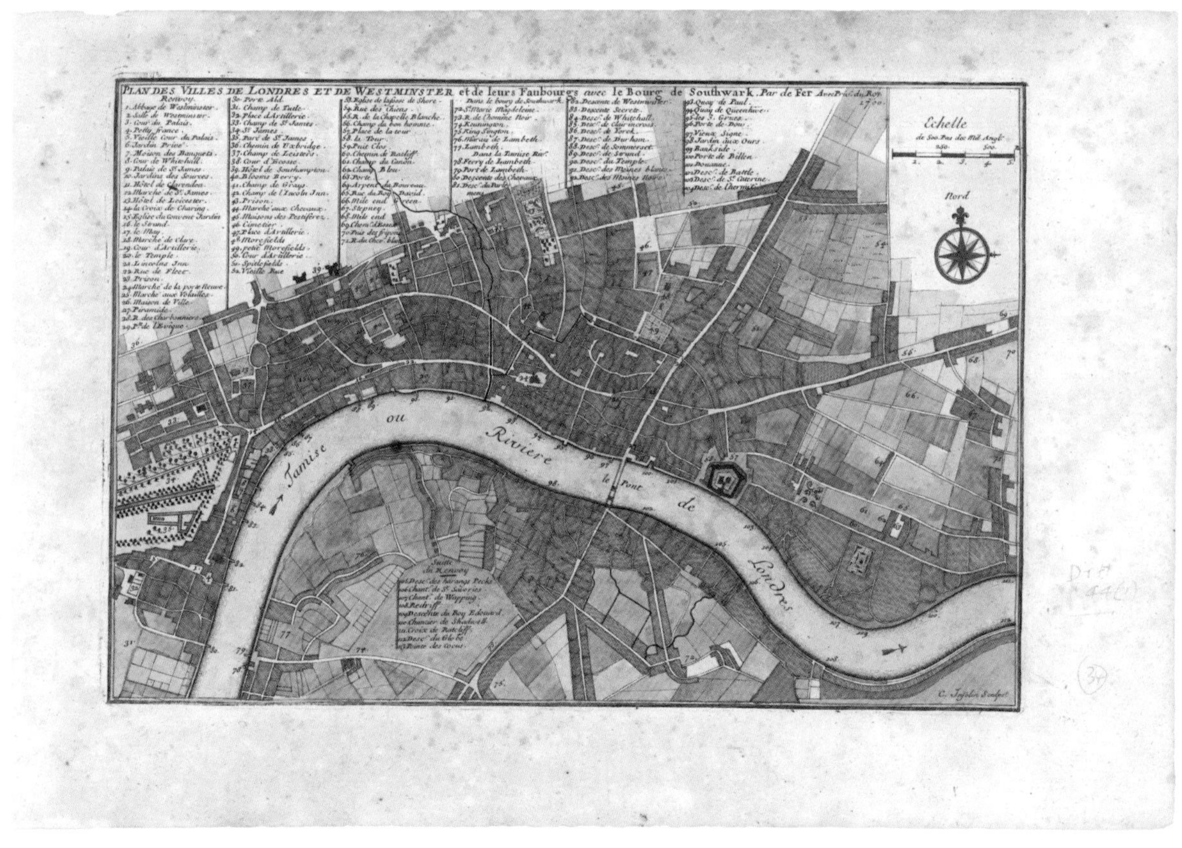

The first cup of tea in England?

22 **MORDEN, Robert and Philip LEA**

London, Westminster and Southwark.

Publication
[London], By Robt. Morden at the Atlas in Cornhill & By Phil. Lea at the Atlas & Hercules in Cheapside, [c. 1700].

Description
Engraved map on two sheets, joined, original outline hand-colour, list of wards, parishes, hospitals, company halls, inns of court, and streets in Westminster.

Dimensions
640 by 1290mm (25.25 by 50.75 inches).

Scale
Approx. 11 inches to 1 staute mile.

References
Howgego 50 (1).

The plan depicts Arlington House, on the site of what is now Buckingham Palace. Arlington House was the residence of Henry Bennet, Earl of Arlington, one of the "Cabal" Ministry, under Charles II. In 1665, the Earl imported from Holland the first pound of tea, at the cost of sixty shillings, therefore making Arlington House, in all probability, the first place in England where a cup of tea was brewed. Arlington House was demolished in 1703 and its site purchased by John Sheffield, Duke of Buckingham. Many other grand residences and important buildings are depicted in bird's-eye view, including Burlington House, Berkeley House, Southampton House, London Bridge replete with houses, the Tower of London, and St Paul's.

Also shown is St. James's Park, originally a low and swampy meadow, belonging to the Hospital for Lepers, which was drained and enclosed, and converted by Henry VIII into a "nursery for deer", charmingly depicted on the present plan.

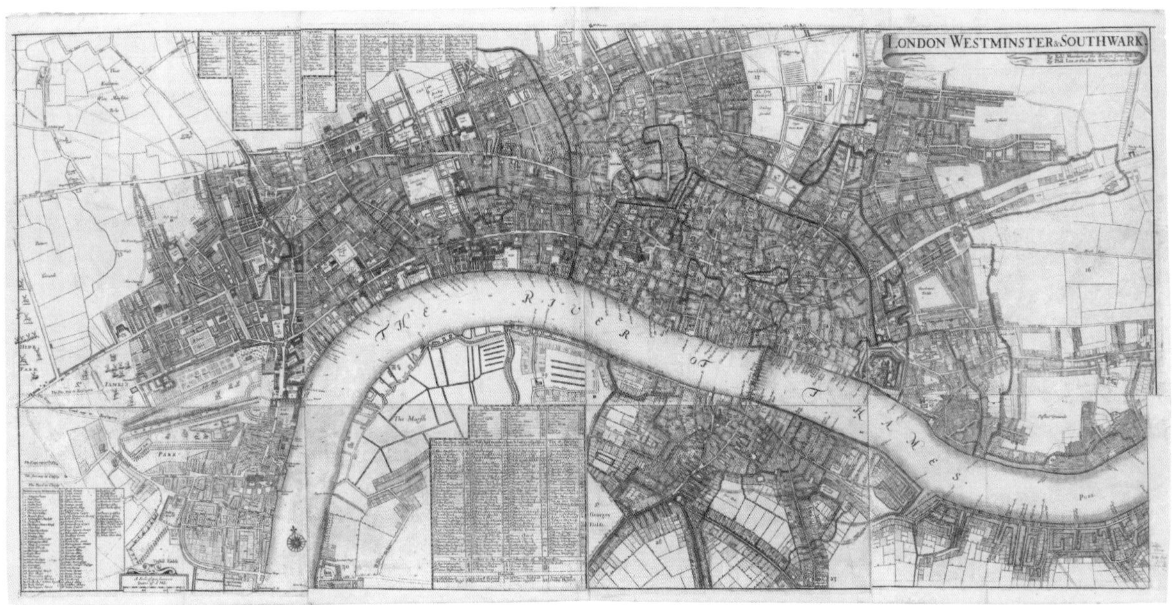

Eastenders

23 **GASCOYNE, Joel**

An Actual Survey of the Parish of St. Dunstan Stepney alias Stebunheath Being one of the Ten Parishes in the County of Middlesex adjacent to the City of London, Describing exactly the Bounds of the Nine Hamlets in ye sd Parish. John Wright Vicar, Charles Walker, William Wheatly, Thomas Walker, Iohn Mumford, Church warden for Ratcliff, Limehous, Popler, Mile end old Town; William Canter, Abram Monfort, William Lee, Humph. Cofter, Church warden for Waping, Spittle Feilds, Bethnal Green, Mile end new Towne.

Publication
London, Taken Ano Dom.1703 by Ioel Gascoyne. Engraved by Iohn Harris, 1703.

Description
Engraved map and separate title, the complete map consists of eight sheets of varying dimensions, the present map lacks sheet [B] (supplied in facsimile), all sheets with good full margins.

Dimensions
1250 by 1150mm (49.25 by 45.25 inches).

Scale
approximately 12 inches to 1 statute mile.

References
BLMC Maps K.Top.28.18.a; For facsimile of the complete map see LTS Publication 150, and accompanying booklet, Joel Gascoyne's Engraved Maps of Stepney, 1702–1704, by William Ravenhill and David J. Johnson (1995).

Extremely rare. A magnificent map extending west to east from Spittlefields to the River Lea and north to south from Bethnal Green to the Isle of Dogs, encompassing the parishes of Ratcliff, Limehouse, Poplar, Mile End Old Town, Wapping, Spittle Fields, Bethnal Green, Mile End New Town, Bromley and Bow.

The map was printed from eight irregular-sized and – shaped copper plates. Curiously, it is not possible to arrange all eight sheets as a rectangle with a continuous engraved border; the title sheet is bordered so as to be located in the upper right corner, and yet the plate for Bow must be placed there for the sake of geographical accuracy. It appears that the author was undecided as to how the finished map should appear and left the final composition up to the owner! In the present example, and in the copy in the King's Collection at the British Library, the title is offered as a separate item. However, the copy in the library of Brasenose College, Oxford, is arranged with the table cut into strips at the sides, with the title (without its rule-border) in the bottom left. The eighth plate – a mere 4" by 6" – lists the public buildings of Bow and in no example we have examined seems to have been placed tidily.

The parallelogram plate provided for the bottom right of the map presented further problems for the engraver in so far as it finished short of where he needed to engrave the rule-border. Harris solves this problem by engraving a neat triangle of border within the margin of the plate to be cut out and pasted on later.

Joel Gascoyne (1650–1705) established himself as a chart-maker, surveyor and cartographer at 'The Sign of the Platt neare Wapping Old Stayres three doares below the Chappell'. As a full member of the Drapers' Company and with apprentices indentured to him, he produced manuscript and engraved charts. The present map was one of his earlier efforts at plotting dry land; a fact that may account for the somewhat unorthodox handling of the task at hand.

Arrangement:

Sheet [A] consists of the title within engraved cartouche carried by putti above an allegorical vignette of the Thames, title surmounted by a portrait of St Dunstan (to whom Stepney Church is dedicated) pinching the devil's nose with a pair of hot tongs (St Dunstan was a goldsmith by profession and so would always have had such implements to hand), engraved table of '...the names of the Courts, Yards, Alleys, Rents and other Remarkable Places in the severall Streets of the respective Hamlets Contained in this Parish' to lower left. This table in fact commences on sheet [B] which is absent from the present collection. Below the cartouche is an explanation of symbols, and two scale-bars. An elaborate compass rose appears in the Greenwich Peninsula on sheet [F].

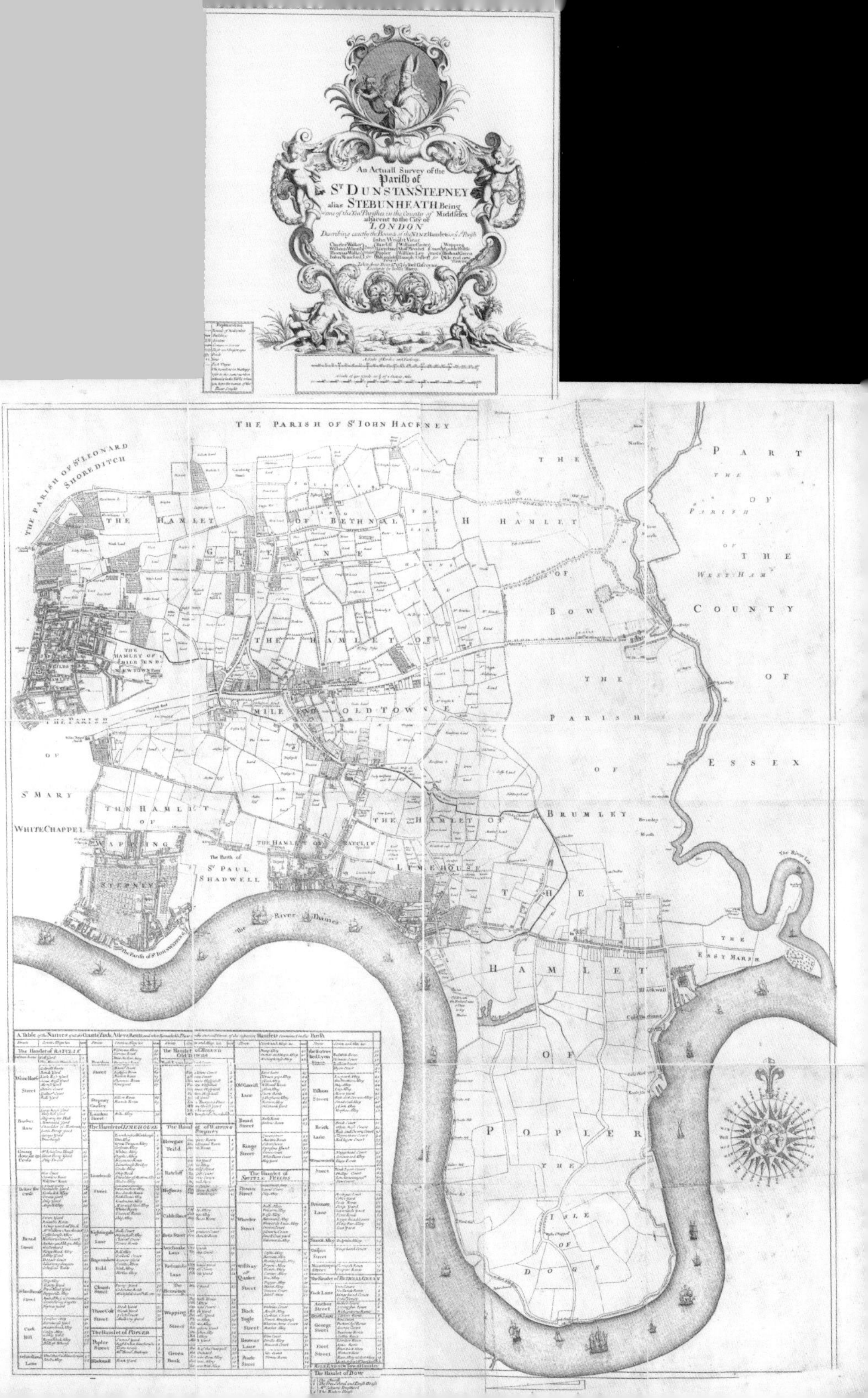

An Actuall Survey of the
Parish of
S^T DUNSTAN STEPNEY
alias STEBUNHEATH Being
one of the Ten Parishes in the County of Middlesex
adjacent to the City of
LONDON
John Wright Vicar
Charles Walker
William Wheatle
Thomas Walker
John Mumford
Ratcliff
Limehouse
Popler
Whitechapel
William Cantey
Abel Mossell
William Lee
Humph. Cullen
Wapping
Spittle Fields
Bethnal Green
Mile end new Towne
Taken Anno Dom 1703 by Joel Gascoyne
Explanation
A Scale of Perches and Furlongs
A Scale of 440 yards or 1/4 of a Statute Mile
THE PARISH OF S^T JOHN HACKNEY
THE PARISH OF S^T LEONARD SHOREDITCH
THE HAMLET OF BETHNAL GREENE
THE HAMLET OF BOW
THE HAMLET OF MILE END NEWTOWN
MILE END OLD TOWNE
THE PARISH OF S^T MARY WHITECHAPEL
THE HAMLET OF WAPPING STEPNEY
THE HAMLET OF RATCLIF
THE HAMLET OF LIMEHOUSE
The Parish of S^T PAUL SHADWELL
The Parish of S^T JOHN WAPPING
THE River Thames
THE HAMLET OF BRUMLEY
Brumley Marsh
PART OF THE PARISH OF WEST HAM COUNTY OF ESSEX
The River Lee
THE EAST MARSH
Blackwall
THE HAMLET OF POPLER
THE ISLE OF DOGS
A Table of the Names of all the Courts, Yards, Alleys, Rents, and other Remarkable Places, in the severall Streets of the respective Hamlets Contained in this Parish
The Hamlet of RATCLIF
The Hamlet of MILE END Old Towne
The Hamlet of LIMEHOUSE
The Hamlet of WAPPING Stepney
The Hamlet of SPITTLE FIELDS
The Hamlet of POPLER
The Hamlet of BETHNAL GREEN
The Hamlet of BOW

Attractive plan and prospect of London

24 **HOMANN, Johannes Baptist**

Accurater Prospect und Grundris der Königl: Gros-Britannisch Haupt und Residentz Stadt London.

Publication
Nuremberg, [c. 1705].

Description
Double-page engraved map, fine original full-wash colour, inset views of the Palace of Whitehall, and the Royal Exchange, prospect of London from Southwark below.

Dimensions
535 by 625mm (21 by 24.5 inches).

References
Howgego 51b.

To the upper left of the plan is a view of the Palace of Whitehall – the royal residence in the seventeenth century; to the upper right is a view of the Royal Exchange, which had been rebuilt by Edward Jarman, after the previous one had been destroyed in the Great Fire. Below left is a list of streets, churches and public buildings. The City of London is highlighted in red. Underneath the plan is a fine prospect of the city listing 63 places of interest.

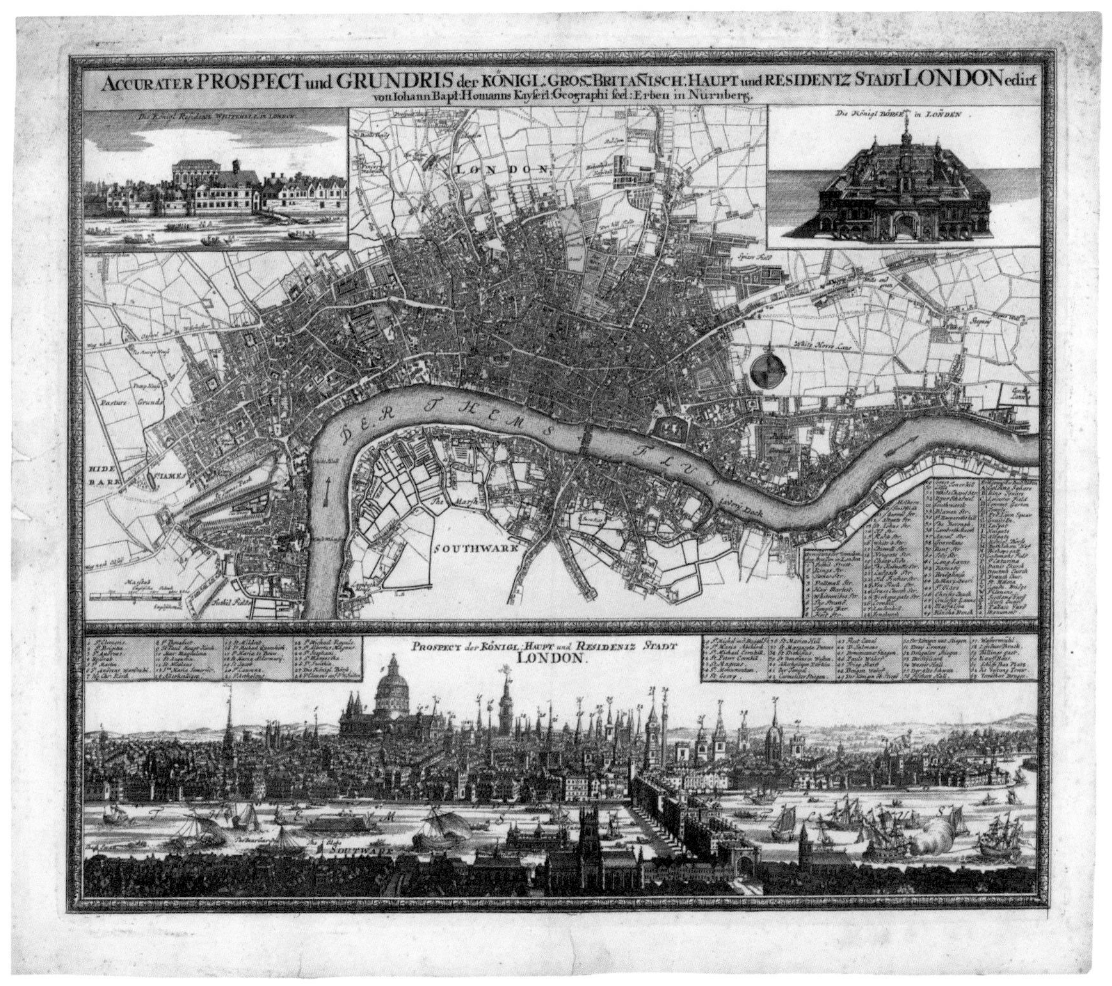

Overton's large and detailed plan

25 **OVERTON, John**

A New Mapp of the Citty of London much Inlarged since the Great Fire in 1666.

Publication
Printed & Sold by John Overton at the White Horse without Newgate London, 1706 [but 1707].

Description
Hand-coloured engraved plan on two sheets joined, extending north to south from Clerkenwell to Southwark, and west to east from St. James's to Whitechapel, some even age toning, and a few old folds reinforced.

Dimensions
635 by 990mm (25 by 39 inches).

Scale
11 inches to 1 statute mile.

References
Howgego 53.

Although based upon Wencelaus Hollar's 1685 plan (Howgego 35), Overton's plan is considerably larger with a great many alterations. It bears the imprint of John Overton's son Henry, who took over the business in 1707. To the upper left are the arms of the City of London, to the upper right a list of churches with and without the City walls; below is a list of the public buildings, streets, taverns, and the location of Henry Overton's shop – marked by the letter 'P' and a white horse. Upon the banks of the Thames, Overton has clearly marked the stairs from which one could hire a ferryman; "Fleet Ditch" or Fleet River is highlighted, as are the Tower of London, Whitehall, Scotland Yard, London Bridge, St James's, Parliament, Lambeth Palace, and St Pauls.

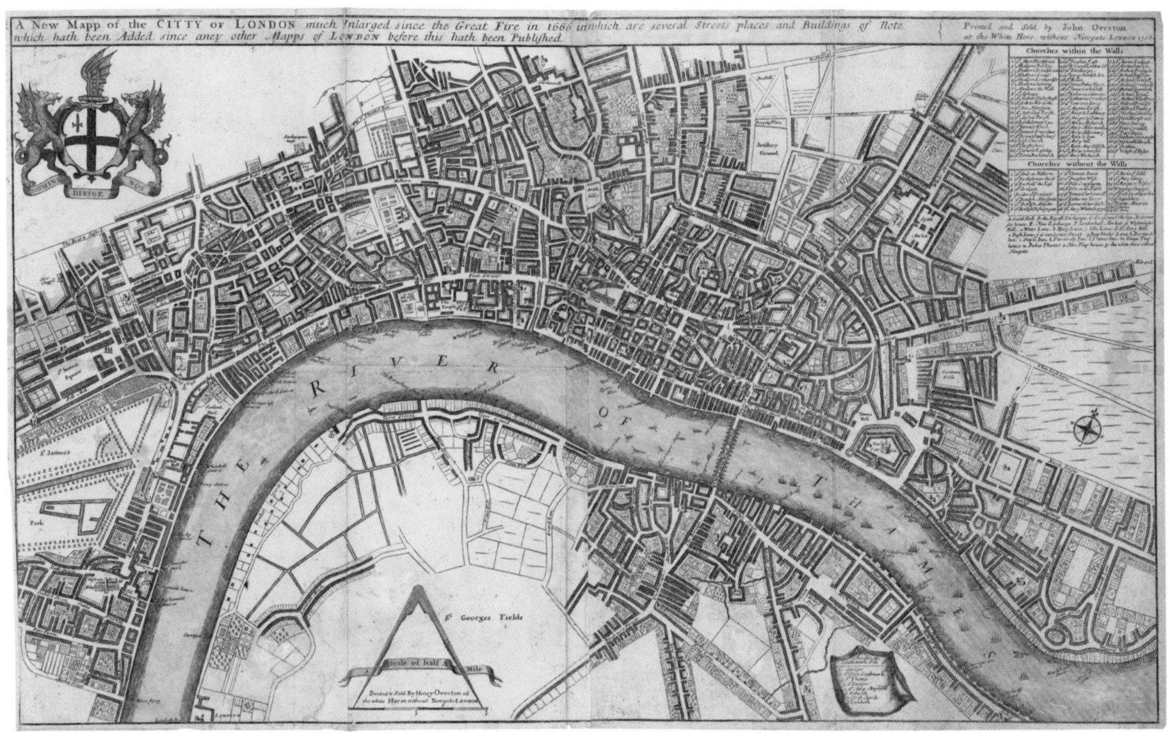

26 OVERTON, John

A New Mapp of the Citty of London much Inlarged since the Great Fire in 1666...

Publication
Printed & Sold by Henry Overton at the White Horse without Newgate, 1706 [but 1707].

Description
Engraved plan on two sheets joined, extending north to south from Clerkenwell to Southwark, and west to east from St. James's to Whitechapel, old joins reinforced.

Dimensions
575 by 950mm (22.75 by 37.5 inches).

Scale
11 inches to 1 statute mile.

References
Howgego 53.

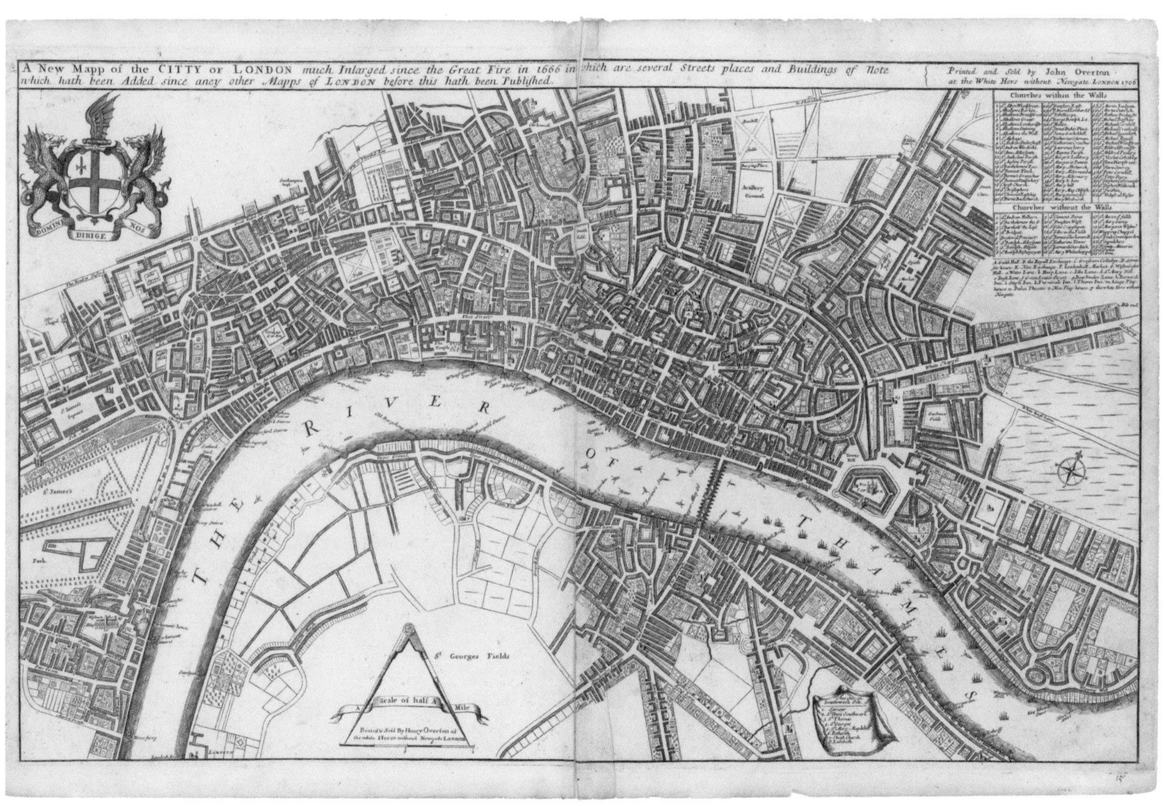

John Seller's rare map of Middlesex

27 **SELLER, John**

The County of Middlesex Actually Survey'd and Delineated By John Seller.

Publication
[London], Sold by Ric. Davies at the 3 Ink Bottles in Castle Alley at the west end of the Royal Exchange, 1710.

Description
Engraved map on two sheets joined, fine original full-wash colour.

Dimensions
635 by 1005mm (25 by 39.5 inches).

Scale
3/4 inch to 1 statute mile.

References
BLMC Maps 3455.(9.); Skelton 115.

The map was first issued by Seller in 1679. The map was to be part of Seller's large folio county atlas of England and Wales, entitled 'Atlas Anglicanus'. However the project, like much of Seller's over-ambitious schemes, never got off the ground, with only six of the counties – Middlesex, Surrey, Hertfordshire, Kent, Buckinghamshire, and Oxfordshire – being surveyed. In 1693, he was forced to sell the plates to Philip Lea, who issued them separately and as part of the composite atlases of England and Wales. After Lea's death in 1700, the business was run by his widow Anne, until her death in 1730. How the map came to bear the imprint of Richard Davies is unclear, however, he was known as a co-publisher of several maps and atlases, and he might have come to some agreement with Anne Lea.

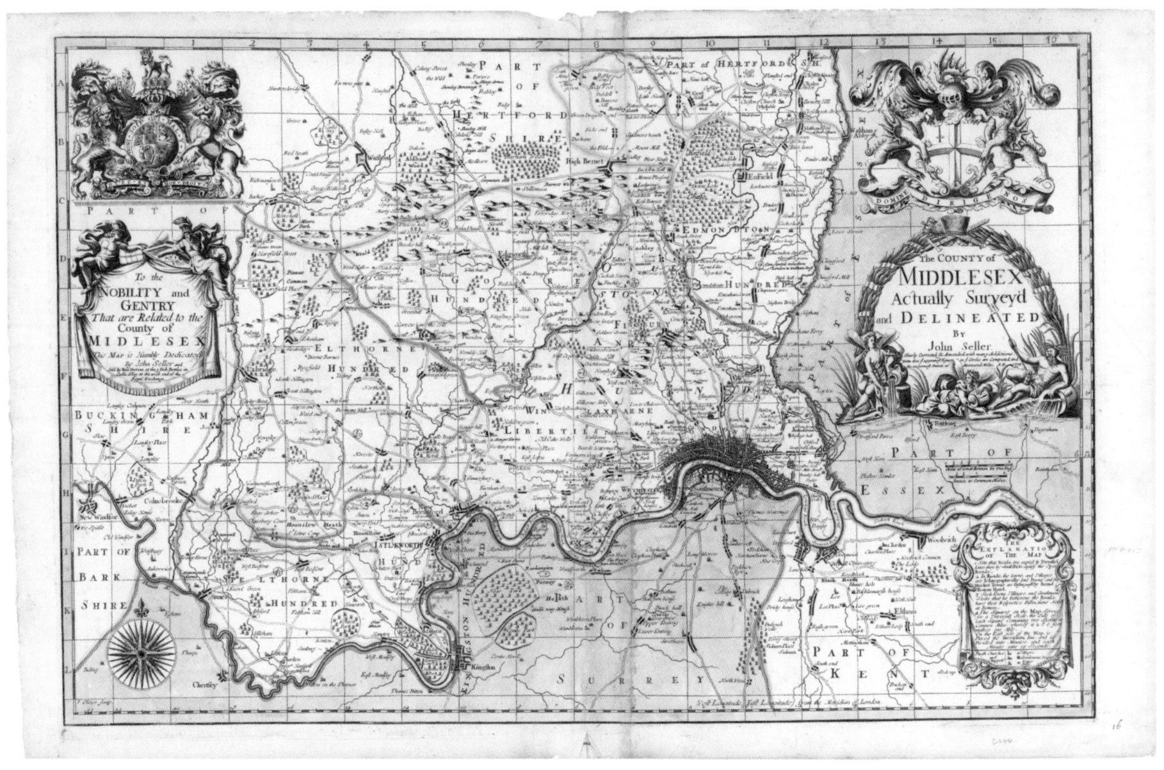

A city of meat eaters

28 **BOWLES, Thomas**

A New and Exact Plan of Ye City of London and suburbs thereof, With the addition of the New Buildings, Churches &c. to this present year... (Not extant in any other) Laid down in such a method that in an Instant may easily be found any Place contain'd therein.

Publication
London, Printed & Sold by Tho: Bowles next to the Chapter-house St. Paul's Church Yard; Sold by John Bowles at the Black Horse in Cornhill 1719.

Description
Engraved map on three sheets, joined, parishes picked out in outline colour, extending north to south from Clerkenwell to Southwark, and west to east from Buckingham House to Stepney, list of streets and squares to left and right margins, trimmed to neatline, a few nicks and tears and some minor loss, skilfully repaired.

Dimensions
610 by 1475mm (24 by 58 inches).

Scale
11 inches to 1 statute mile.

References
Howgego 63 (1).

Large and detailed plan of London.

The map bears grid lines for ease of reference, the major public buildings such as Westminster Abbey, the Houses of Parliament, St Paul's Cathedral, Buckingham House, and the Tower of London are represented as bird's-eye views.

The text upon the map gives the extent of London as "7500 Geometric paces, that is above 7 English Miles and a half" from west to east and "2500 paces, or 2 miles and a half" from north to south. There are about 5000 streets. The number of houses is calculated to be 110,000.

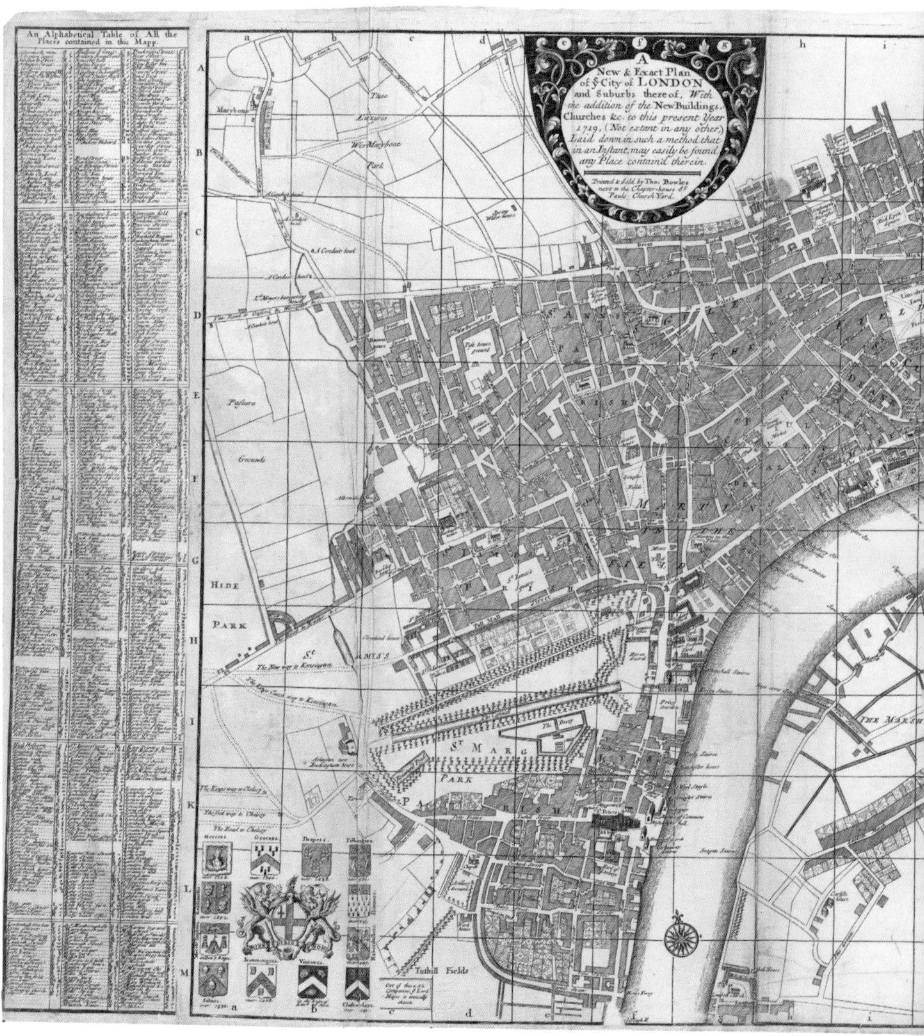

The population is guessed at "by what is eaten": "There were in one Year, when it was less by two thirds [ie. in the mid seventeenth century], 67500 Beefs, 10 times as many Sheep, Besides Poultry &c. also every Year is brought into the River 400,000 at least Charldon of Coales". Also the number of "Buryings" was said to number 26,000 per year. The amount of ale and beer produced is said to be 20,000 barrels. Bowles acknowledges the source of these numbers as Sir William Petty (1623–1687), who through some crude statistical analysis calculated the population in the mid-seventeenth century. He estimated that there were 115,846 families and 695,076 souls; more than Paris, Rome, Rouen or Amsterdam, which were said to have some 500,000 souls. To the left of the text is a list of watermen and hackney carriage rates.

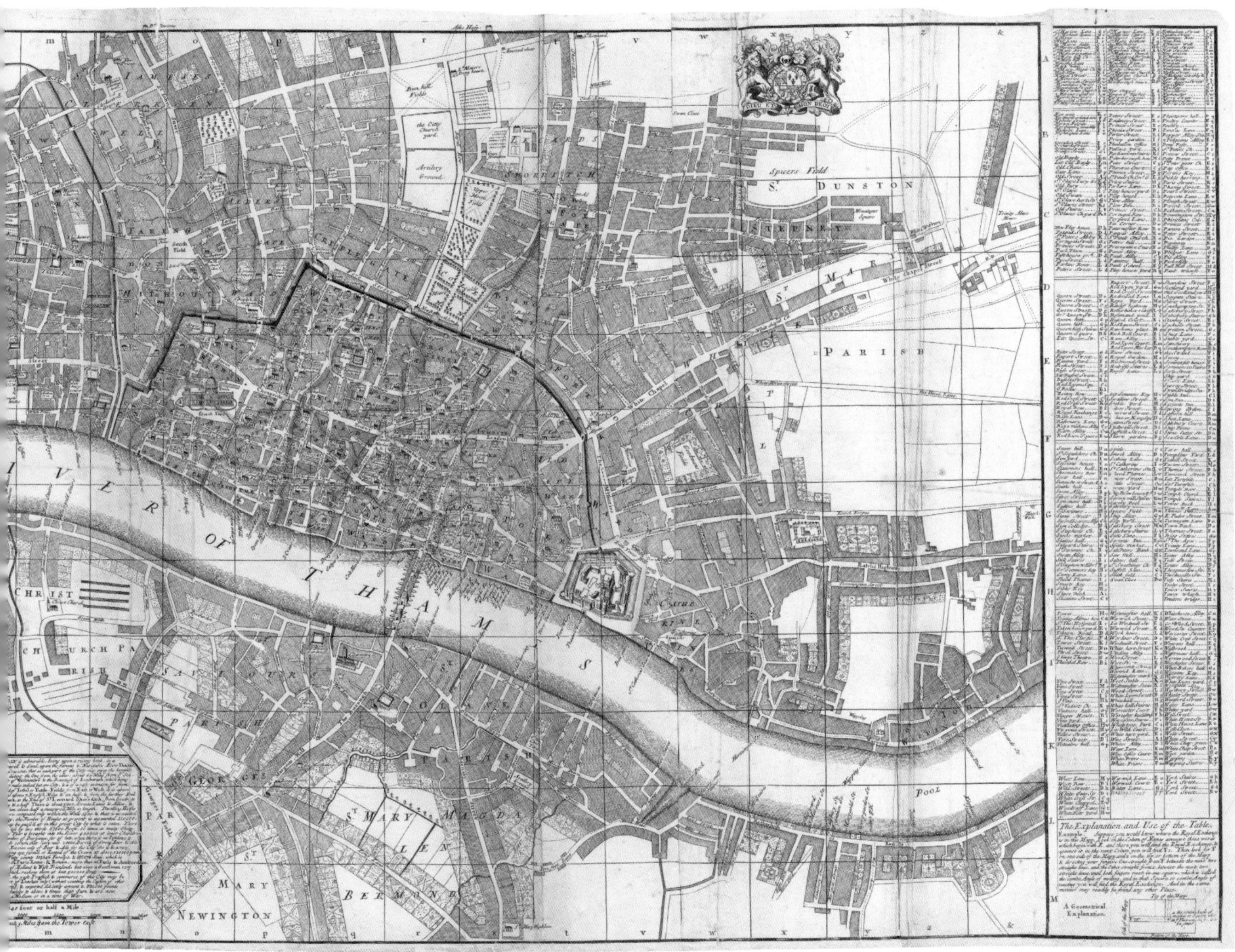

Three states of Bowles' Map

29 **BOWLES, Thomas**

A New and Exact Plan of Ye City of London...

Publication
London, Printed & Sold by Tho: Bowles next to the Chapter-house St. Paul's Church Yard, 1723.

Description
Engraved plan on three sheets joined, original outline hand-colour, a few marks and a few small tears skilfully repaired.

Dimensions
620 by 1255mm. (24.5 by 49.5 inches).

Scale
11 inches to 1 statute mile.

References
Howgego 63 intermediate state between 3 and 3a.

The present plan is an intermediate state between Howgego 63 (3) and (3a). The date in the title is the same as state 3, however, an extra sheet has been added to the extend the plan to the east to include St Dunstan and Stepney; yet the plan does not include the areas of Mile End and Limehouse present in state 3a.

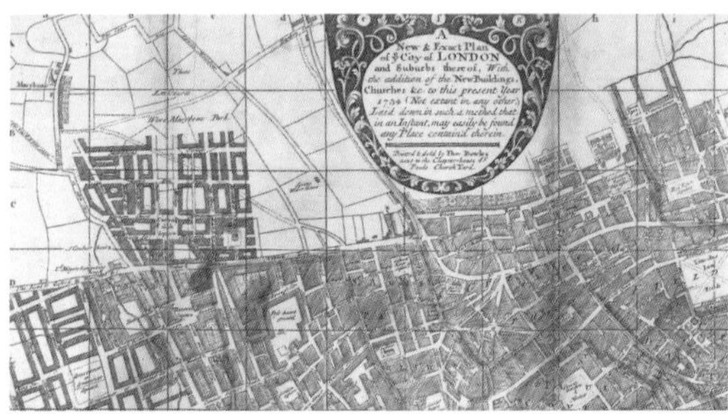

Item 29 (detail above)
Item 30 (detail below): The development of Cavendish Square and Marylebone
Item 31 (right): The map extended eastwards

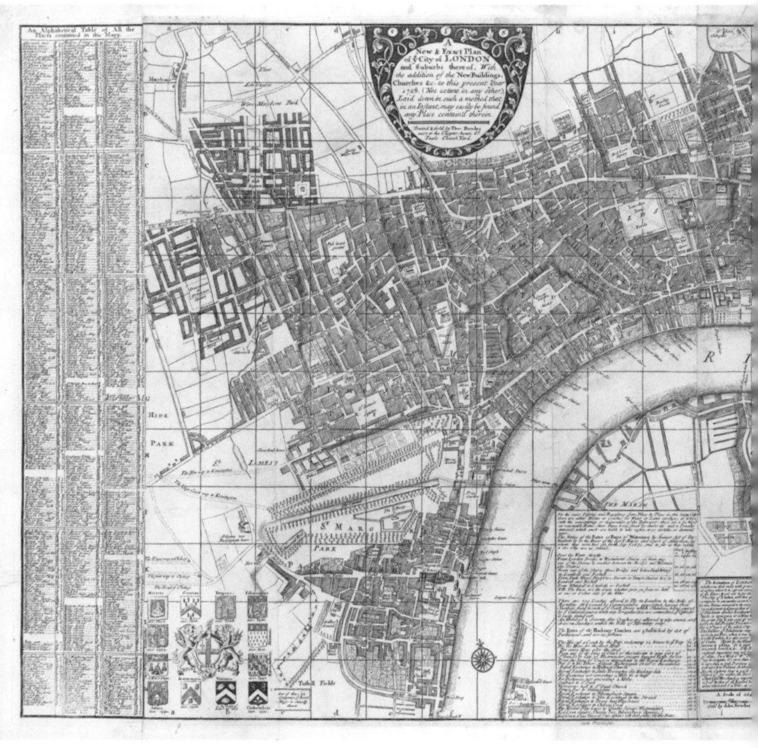

30 **BOWLES, Thomas**

A New and Exact Plan of Ye City of London...

Publication
[London], Printed & Sold by Tho: Bowles next to the Chapter-house, St. Paul's Church Yard, 1728.

Description
Engraved plan, on two sheets joined, tear to join at foot not affecting image.

Dimensions
620 by 1255mm. (24.5 by 49.5 inches).

Scale
7½ inches to 1 statute mile.

References
Howgego 63 intermediate between states 4 and 5.

31 **BOWLES, Thomas**

A New and Exact Plan of Ye City of London...

Publication
London, Printed & Sold by Tho: Bowles next to the Chapter-house St. Paul's Church Yard. Sold by John Bowles at the Black Horse in Cornhill London, 1734.

Description
Engraved plan, on five sheets joined, fine original outline hand-colour, list of places upon the plan to left and right borders, slight discolouration to folds.

Dimensions
610 by 1470mm (24 by 57.75 inches).

Scale
11 inches to 1 statute mile.

References
Howgego 63 (6)

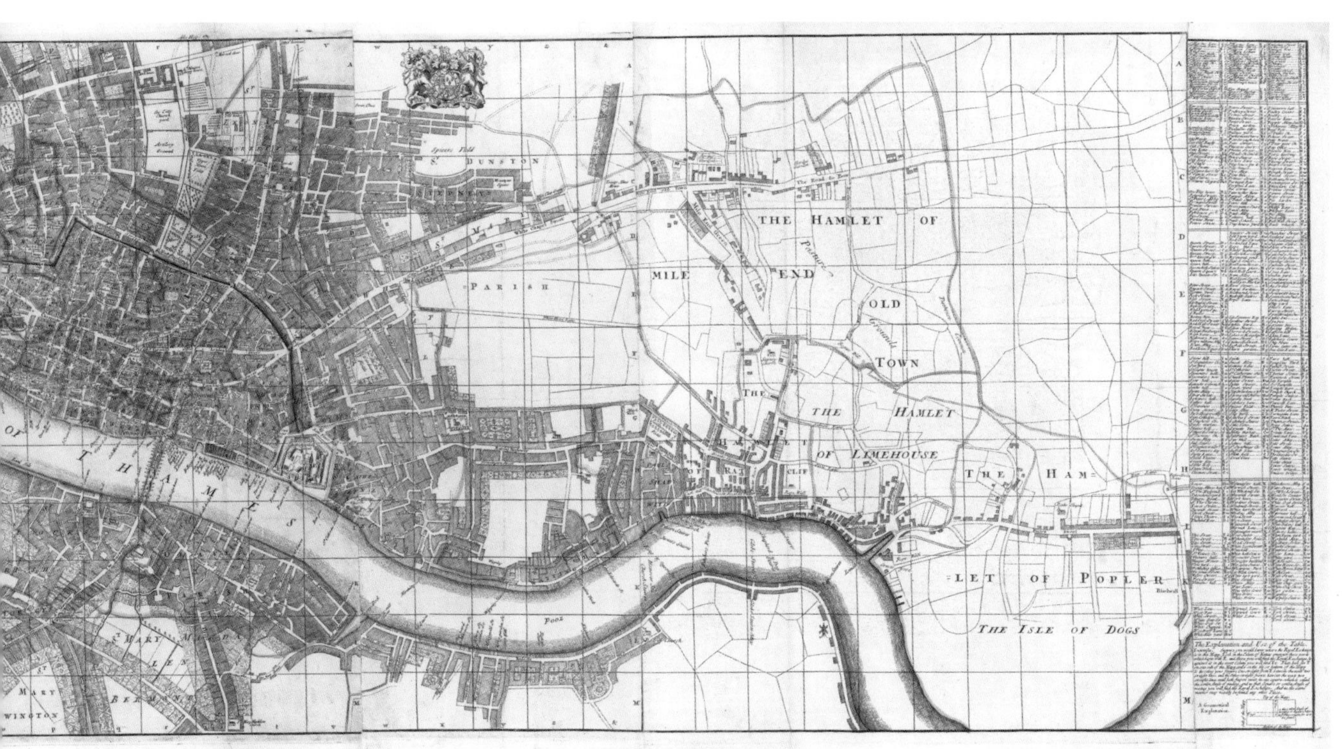

Dedicated to the Governor of the Bank of England

32 **PARKER, S.**

A Plan of the City's of London, Westminster and Borough of Southwark with the new Additional Buildings; Anno 1720.

Publication
London, Revised by John Senex, 1720.

Description
Double-page engraved plan.

Dimensions
530 by 640mm (20.75 by 25.25 inches).

Scale
5½ inches to 1 statute mile.

References
Howgego 65.

The plan is dedicated to Sir Peter Delme, a wealthy merchant who had been knighted in 1714. Delme served as Governor of the Bank of England between 1715–1717, and was made Lord Mayor of London in 1723. His name appears in an elaborate cartouche surrounded by the City's heraldic dragons, the City's regalia, together with symbols of trade, art, and justice. Below the plan is a reference table listing churches, public buildings, parish churches within the walls, parishes part within and part out of the freedom, out parishes, and names of the wards of the City of London.

The plan was published in a 'A New General Atlas Containing a Geographical and Historical Account of the World', published in 1721.

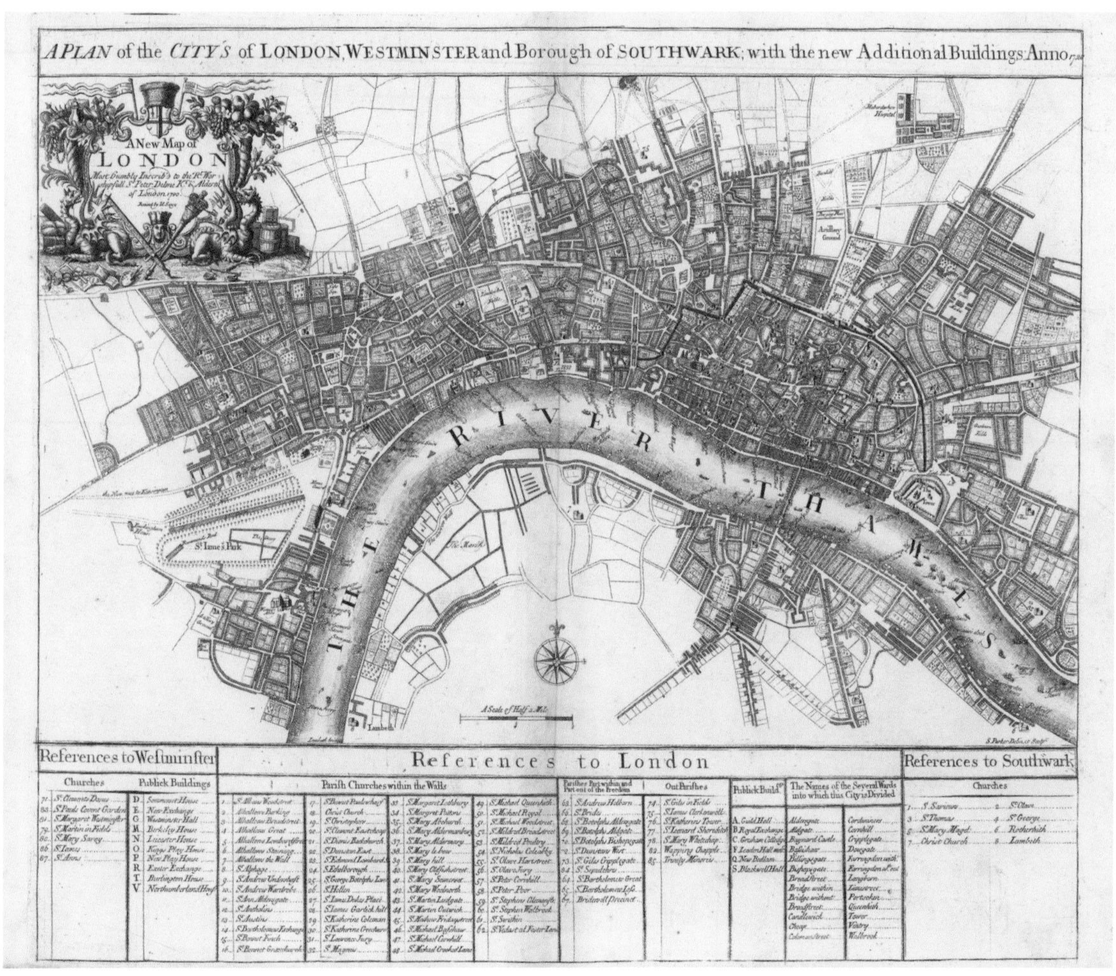

Early eighteenth century London

33 **SMITH, Joseph**

A New and Exact Plan of the Cities of London & Westminster and the Borough of Southwark with all ye Additional New Buildings to ye Present Year 1725.

Publication
London, Printed for & Sold by I. Smith, Print & Map Seller next ye Fountain Tavern in ye Strand, 1725.

Description
Engraved plan, on two sheets joined.

Dimensions
630 by 1030mm (24.75 by 40.5 inches).

Scale
10 inches to 1 statute mile.

References
Howgego 71 (2).

Large and detailed plan of early eighteenth century London.

The plan bears grid lines for ease of reference; the title appears to the right of the plan in a fine cartouche. Many of the most prominent public buildings are shown in elevation, including Lincoln's Inn, St Paul's, Temple, Somerset House, and Burlington House. The plan shows much of the new West End development including Grosvenor Square, Mount Street, and Cavendish Square. Below the plan is a description of London, watermen and hackney coach rates, and a list of important public buildings and streets with their grid reference. The rates for watermen are given for both "oars" (a two man boat) or Sculls (single man); the hackney cabs are said to number 800 and are allowed to "Ply in London", and "Bills of Mortality", all licenced cabs bear a number on the coach door.

This edition appears in Joseph Smith's edition of the 'Nouveau Theatre de la Grande Bretagne', Vol. III, the title page of which bears the date 1724.

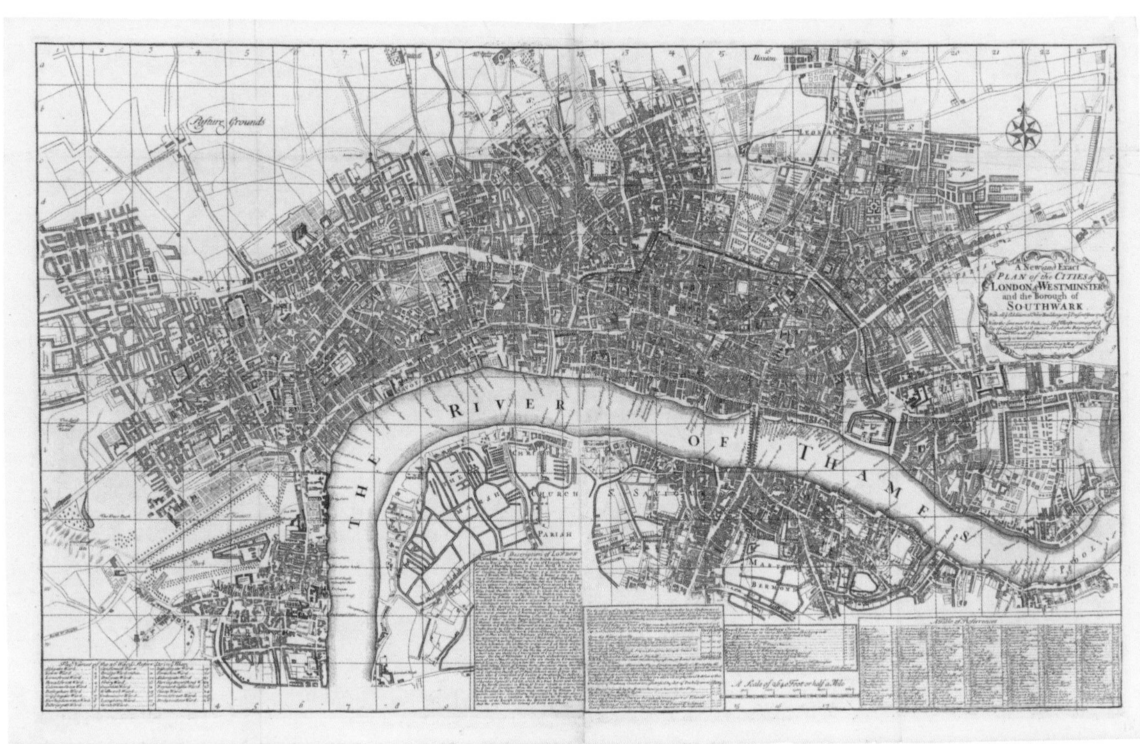

A scarce French plan

34 **DANET, Guillaume**

Plan de la Ville de Londres et de Westminster avec le Bourg de Soutwark, leurs Faubourgs et leurs Environs.

Publication
A Paris, chez Danet sur le Pont Notre Dame a la Sphere Royalle, 1727.

Description
Engraved plan, inset views of Monument and Mary-le-Bow church, tear to old fold in upper margin skilfully repaired.

Dimensions
582 by 810mm (23 by 32 inches).

Scale
5 inches to 1 statute mile.

References
Howgego 73.

Attractive plan of London.

To the left and right of the title is an explanation in French of London and its principal buildings. Below the plan are two views, one of Monument – the great doric column designed by Wren and Hooke to commemorate the Great Fire; and the other, St Mary-le-Bow church; another Wren masterpiece that was completed in 1680 and considered at the time the second most important church in London, after St Paul's. Also below the plan is a list of the most important streets, hospitals, churches, public buildings, and gateways. The plan bears grid lines for easy reference.

Guillaume Danet was the son-in-law of the well-known cartographer Nicholas de Fer. However, Danet was far less prolific a mapmaker, and consequently his maps are rare.

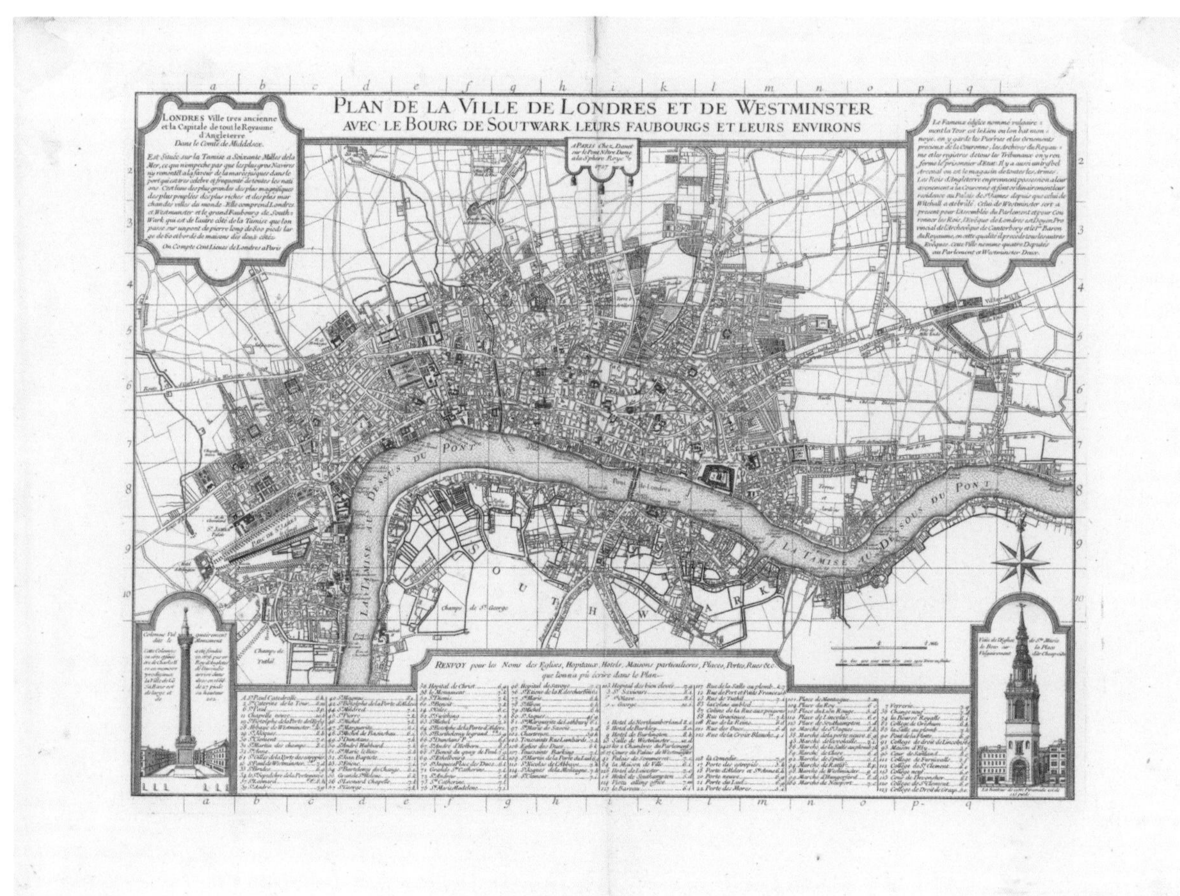

Morgan's map lives on...

35 **JEFFERYS, Thomas**

New and Exact Plan of the City's of London and Westminster and the Borough of Southwark And the Additional New Buildings Churches &c to the present Year 1735.

Publication
London, Printed and Sold by Thos. Jefferys Geographer to His Royal Highness the Prince of Wales in Red Lyon Street near St. John's Gate, 1735.

Description
Engraved plan, on six sheets joined.

Dimensions
855 by 1210mm (33.75 by 47.75 inches).

Scale
11¾ inches to 1 statute mile.

References
Howgego 79.

Large and detailed plan of early eighteenth century London.

The plan bears grid lines for ease of reference; the title appears below the plan in a fine cartouche. Many of the most prominent public buildings are shown in elevation, including Lincoln's Inn, St Paul's, Temple, Somerset House, and Burlington House. The plan shows much of the new West End development including Grosvenor Square, Mount Street, and Cavendish Square. Below the plan is a description of London, watermen and hackney coach rates, and a list of important public buildings and streets with their grid reference. The rates for watermen are given for both "oars" (a two man boat) or Sculls (single man); the hackney cabs are said to number 800 and are allowed to "Ply in London", and "Bills of Mortality", all licenced cabs bear a number on the coach door.

Thomas Jefferys was in business as a map seller, first in Red Lyon Street, Clerkenwell, and then in St. Martin's Lane, later going into business with William Faden. In 1746, Jefferys was appointed Geographer to Frederick, Prince of Wales, and in 1757, Geographer to George, Prince of Wales, later George III. Jefferys republished Morden and Lea's version of William Morgan's map in 1732 and in 1735 published this map, which, though on a reduced scale, is largely based on that map.

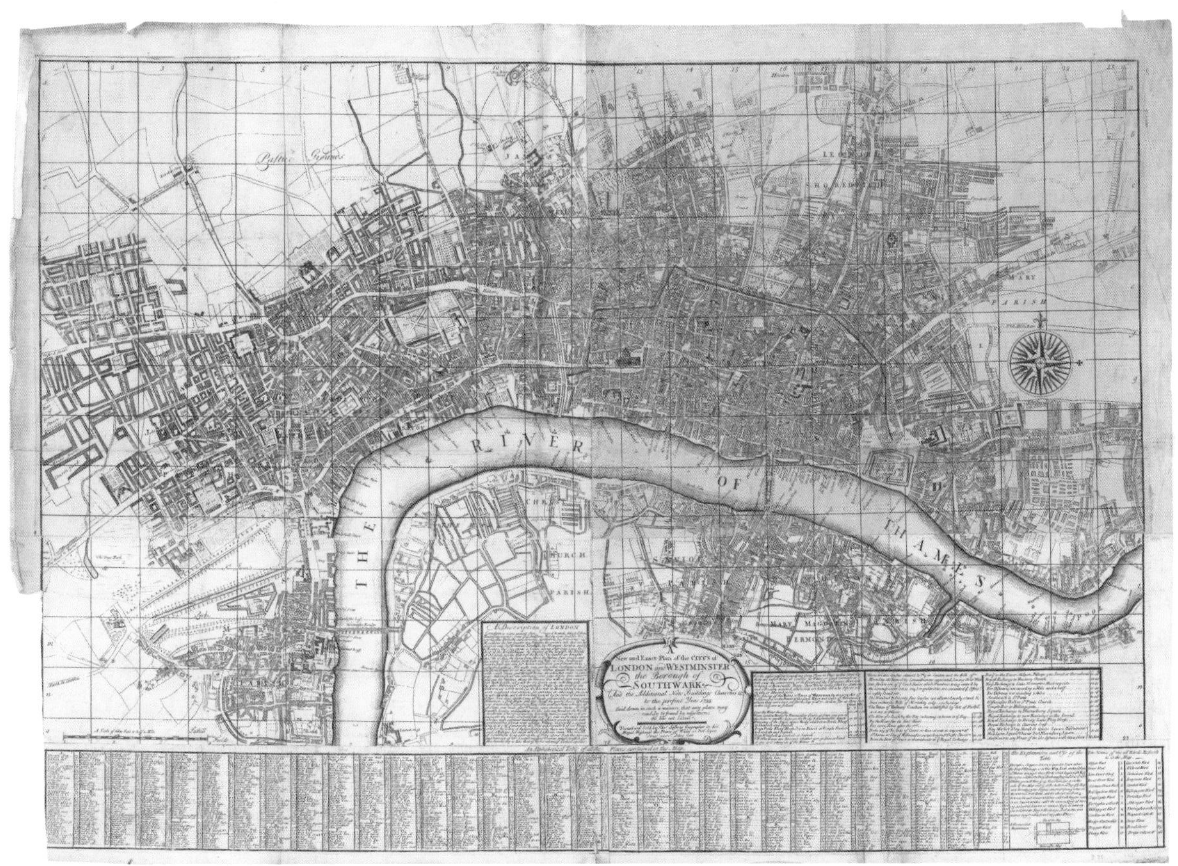

"Westminster New Bridge"

36 **BOWLES, John**

London Surveyed, or A New Map of the Cities of London and Westminster and the Borough of Southwark Shewing the several Streets and Lancs with most of ye Alleys & Thorough Fares: with the additional new Buildings to this present Year 1738.

Publication
London, Printed for John Bowles at the Black Horse in Cornhill, 1738.

Description
Engraved plan, on five sheets joined, list of places upon the plan to left and right border, inset views of the Banqueting House, the Treasury, the Royal Exchange, St Paul's, Monument, Bank of England, prospect of London from Southwark, some minor wear to folds.

Dimensions
630 by 1450mm (24.75 by 57 inches).

Scale
11 inches to 1 statute mile.

References
Howgego 80, between editions (1) and (2).

The elaborate title cartouche is copied from Parker's 'A Plan of the City's of London...' of 1720. The plan bears grid lines for easy reference, with many of the most important public buildings shown in elevation. The Thames is shown teeming with barges, watermen, and to the east of London Bridge; sailing ships. Also shown is "Westminster New Bridge", however, not with thc dcsign that would be adopted later.

There are also views of Banqueting House, the Treasury, the Royal Exchange, St Paul's, Monument, the Bank of England, and a prospect of London from Southwark. To the lower left is a table listing the rates of watermen and coachmen, and to the left and right borders a list of places to be found upon the plan together with their grid reference.

The map is an intermediate state between Howgego 80 states (1) – bearing a date of 1736 – and (2) bearing a date of 1742. The date on the present map has been altered to 1738.

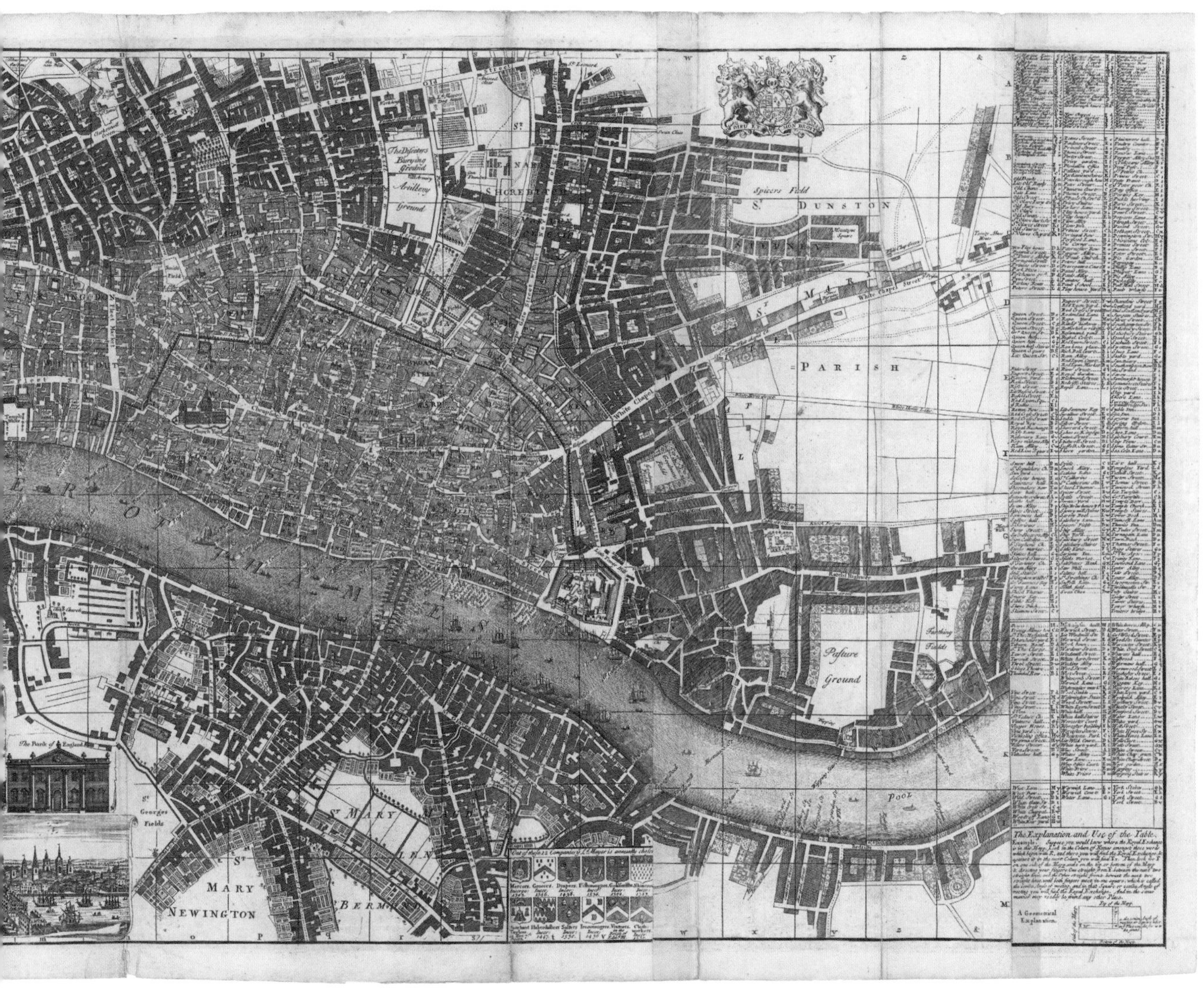

A river runs through it

37 **HOMANN, Heirs**

Vrbium Londini et West-Monasterii nec non, Suburbii Southwark accurata Ichnographia...usque ad A. 1736.

Publication
Nuremberg, Homann Heirs, 1736.

Description
Engraved plan on three sheets, fine original hand-colour, extending north to south from Shoreditch to Newington, and west to east from Buckingham Palace to the Bow Road, inset views of St. Paul's Cathedral, St. James's Square, Custom House, and the Royal Exchange.

Dimensions
(if joined) 500 by 1700mm (19.75 by 67 inches).

Scale
9½ inches to 1 statute mile.

References
Howgego 81.

The plan is finely coloured to highlight: the City of London with its wards in and outside the city walls; the City of Westminster; and Southwark. A great deal of traffic is depicted on the Thames – to emphasise the importance of the river to London – these include barges, water boatmen, and, to the east of London Bridge, sailing vessels. The importance of maritime trade and affairs is also shown in the view of the Custom House, which had been rebuilt some nine years earlier by Thomas Ripley. The title in German and Latin is housed within an elaborate cartouche, which is surmounted by the royal coat-of-arms. To the title's right is a key giving the German translation of English terms used upon the plan.

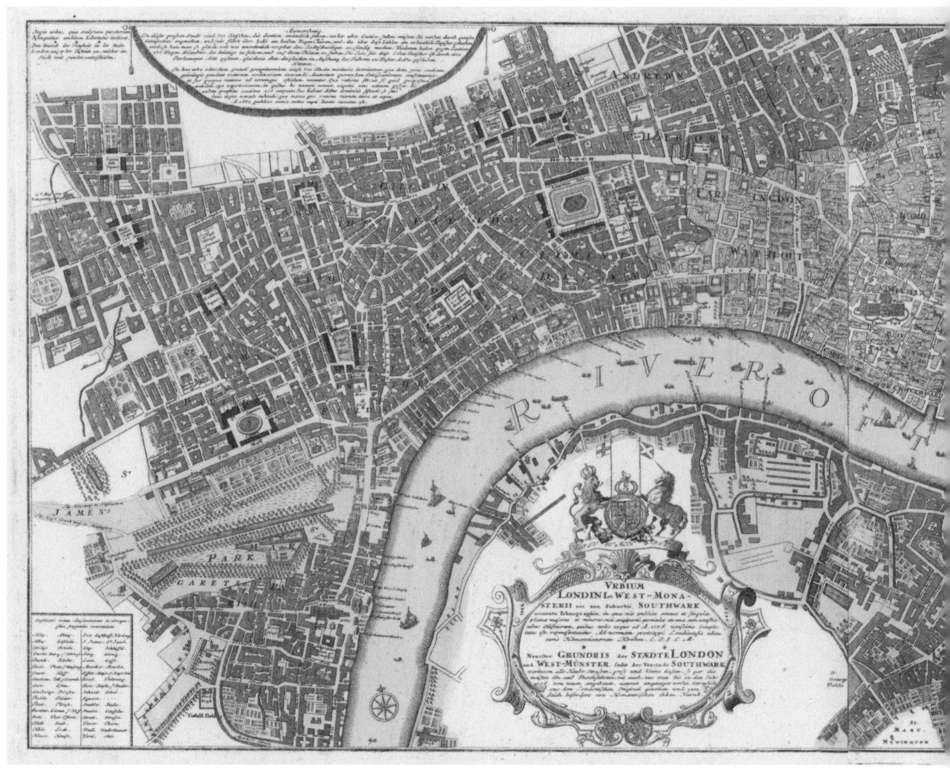

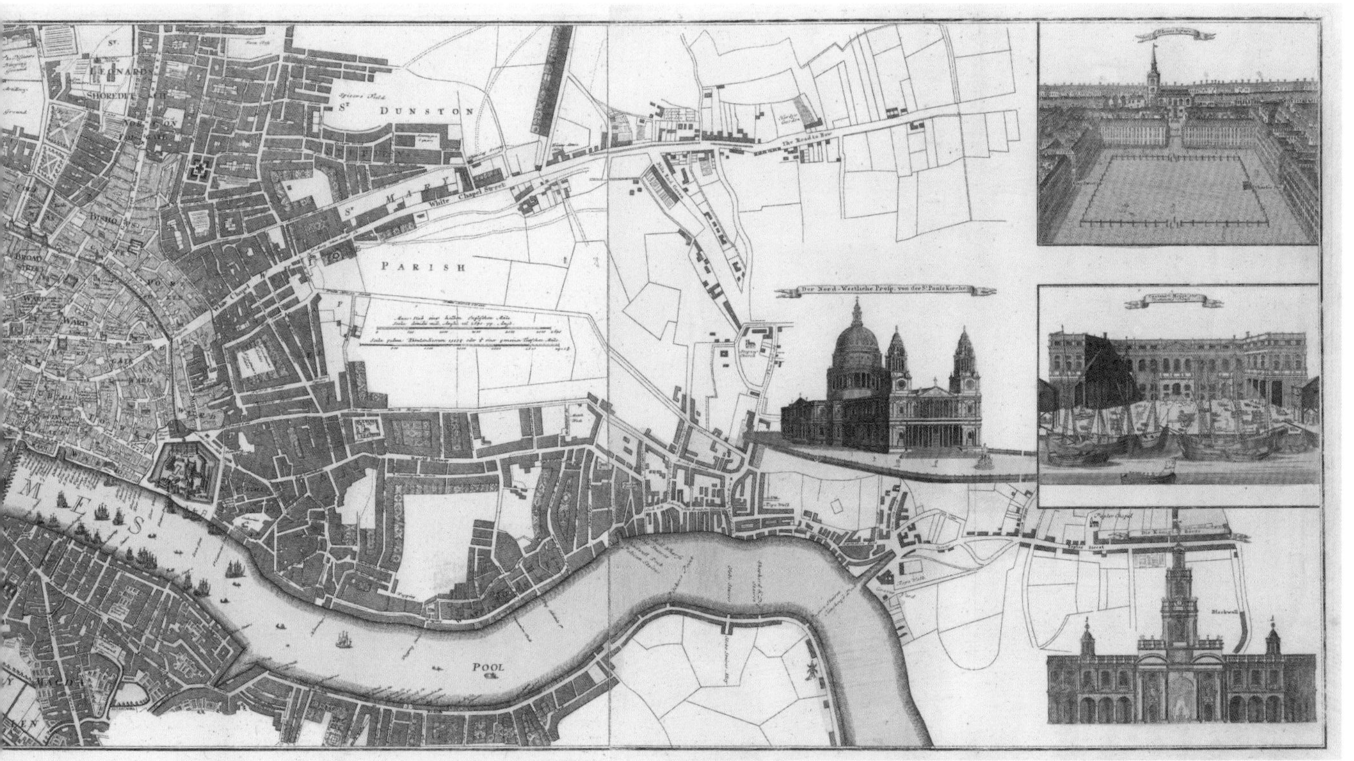

Sᵀ DUNSTON
PARISH
POOL

The Great Livery Companies

38 **FOSTER, George**

A New and Exact Plan of the Cities of London and Westminster & the Borough of Southwark to this present Year 1738.

Publication
London, Printed & Published according to Act of Parliament Aug. 30 1738 and Sold by Geo: Foster at the white Horse, St. Paul's Churchyard, 1738.

Description
Engraved plan on two sheets joined, original outline hand-colour, trimmed to neatline, a few nicks and tears, with some loss, skilfully repaired.

Dimensions
570 by 1020mm (22.5 by 40.25 inches).

Scale
7½ inches to one mile.

References
Howgego 82 (1).

A fine plan of London.

The plan is dedicated by George Forster to John Barnard the Lord Mayor of London, and the MPs for London. To the left of the dedication are the arms of the 12 Great Livery Companies. Many of the most important public buildings are shown in elevation, such as Buckingham Palace, St James's Palace, and St Paul's. The Liberties of the City of London are marked by stippling, whilst those of Southwark and Westminster by hatchures. Westminster Bridge is shown but not with the design that would be adopted when it opened in 1750. Below the plan is a description of London, together with the rates and fares of the hackney coachmen and watermen, and a table referring to the various public buildings and churches upon the plan with their grid reference.

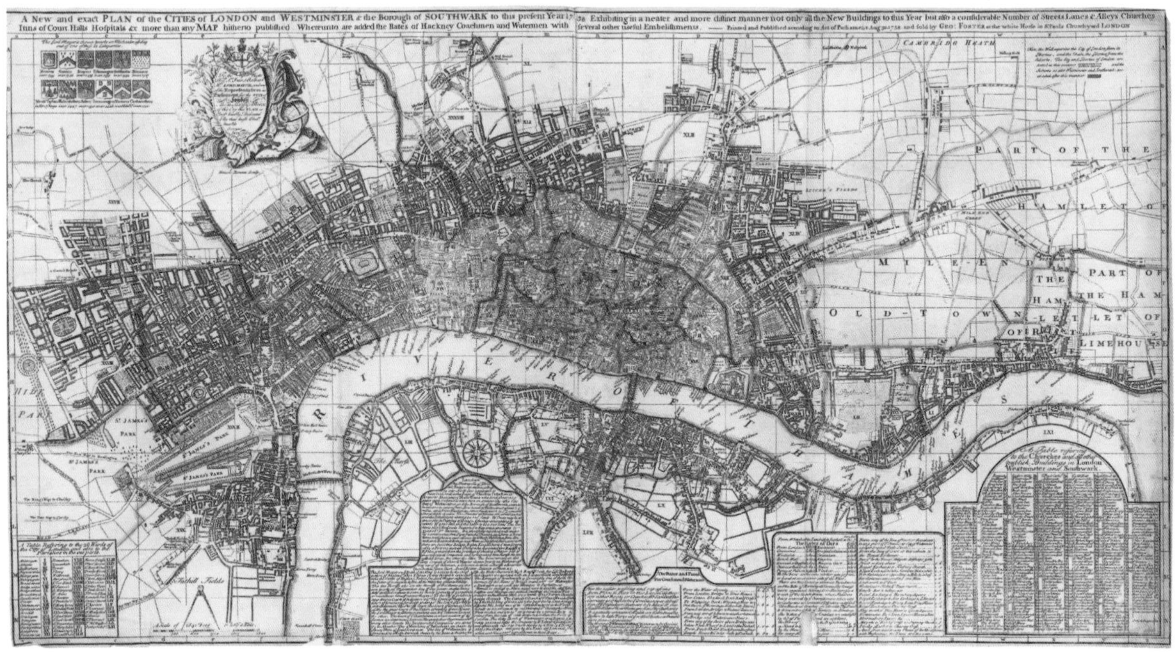

A fine prospect of London and Westminster

39 **HOMANN, Heirs**

Regionis, quae est circa Londinum...

Publication
[Nuremberg], 1741.

Description
Engraved map, original hand-colour, prospect of London and Westminster below.

Dimensions
530 by 605mm (20.75 by 23.75 inches).

Scale
approx. 2½ inches to 1 statute mile.

References
Howgego 88.

This finely engraved map was, as Homann acknowledges in the title, based upon Thomas Bowles' map of a year earlier: 'A New Correct Map of thirty Miles Round London...'. The map is coloured to show hundreds and counties, and depicts towns that return MPs, market towns, churches, gentlemen's houses, castles, windmills, watermills, and villages. Below the map is a prospect of London and Westminster, which has been reduced from a two-sheet 'South Prospect of London and Westminster', printed and sold by John Bowles, circa 1722.

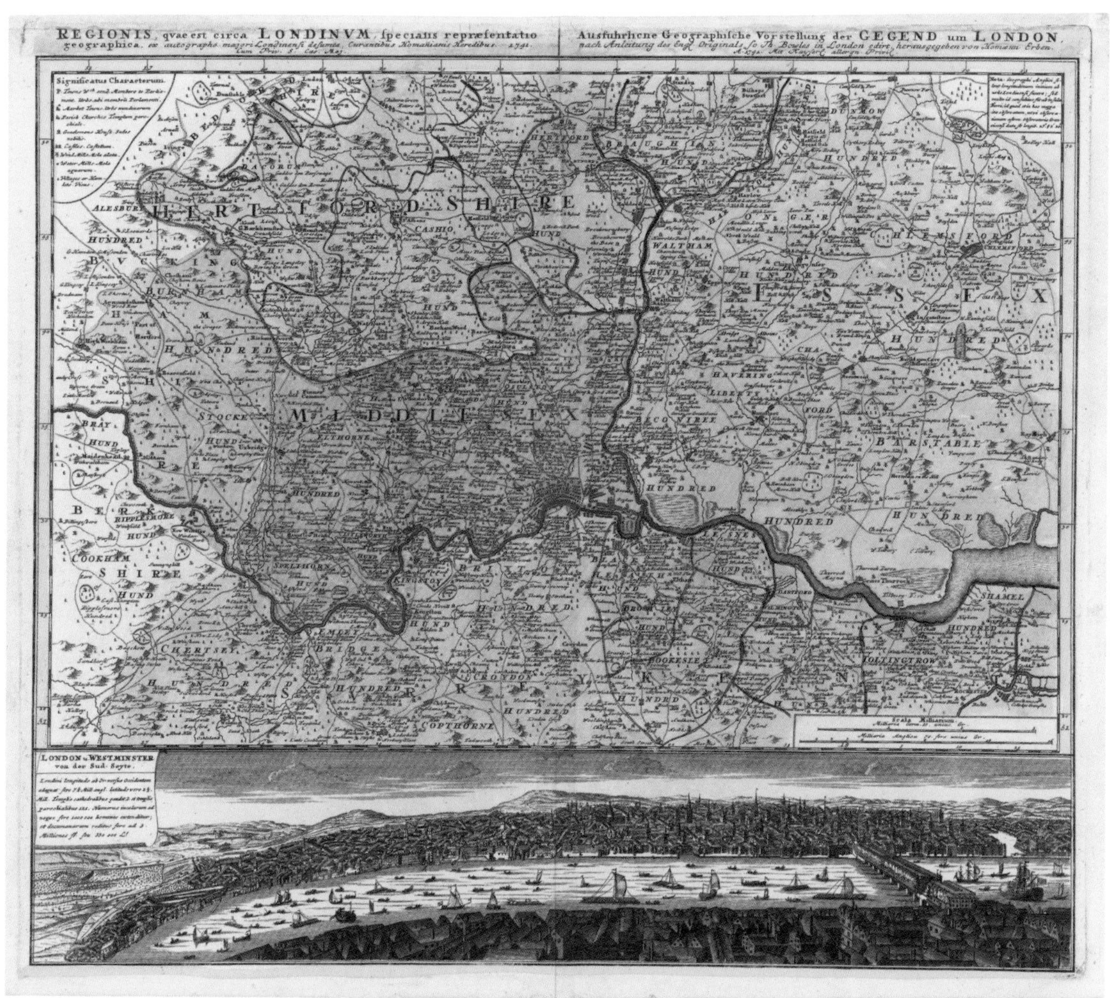

John Rocque's magnificent map of Georgian London

40 **ROCQUE, John**

A Plan of the Cities of London and Westminster and Borough of Southwark, with the contiguous Buildings. From an Actual Survey taken by John Rocque, Land Surveyor and engraved by John Pine.

Publication
London, John Pine and John Tinney, 1746.

Description
Engraved plan on 24 sheets, joined, and mounted on linen.

Dimensions
(approx.) 2025 by 3840mm
(79.75 by 151.25 inches).

References
Howgego 96 (1)

John Rocque, a French Huguenot, emigrated with the rest of his family to London in the 1730s, where he began to ply his trade as a surveyor of gentleman's estates, and with plans of Kensington Gardens, and Hampton Court. However, in 1737 he applied his surveying skills to a much great task, that of surveying the entire built-up area of London. Begun in the March of 1737, upon a scale of 26 inches to 1 statute mile, the map would take nine years to produce, eventually being engraved upon 24 sheets of copper and published in 1746. The plan stretches west to east from Hyde Park to Limehouse and north to south from New River Head to Walworth.

TOTTENHAM COURT ROAD
Timber Yard
MORTIMER STREET
MARGARET STREET
CASTLE STREET
OXFORD STREET
BROAD ST GILES'S
HIGH HOLBOURN
GREAT RUSSELL STREET
BLOOMSBURY SQUARE
RED LYON SQUARE
GREAT ORMOND STREET
THEOBALDS ROW
BEDFORD ROW
EAGLE STREET
HART STREET
LINCOLN'S INN FIELDS
PORTUGAL ROW
GREAT QUEEN STREET
LONG ACRE
Covent Garden Market
SOHO SQUARE
GOLDEN SQUARE
LEICESTER FIELDS
HANOVER SQUARE
PRINCES STR.
GR. MARLBOROUGH STR.
BROAD STREET
CONDUIT STREET
GREAT GEORGE STREET
SAVIL ROW
CLIFFORD STR.
EVANS ROW
HAY HILL
OLD BOND STREET
DOVER STREET
ALBEMARLE STREET
BERKLEY STREET
PICCADILLY
COVENTRY STR.
THE HAY MARKET
CHARLES STREET
ST JAMES'S SQUARE
ST JAMES'S STREET
KING STREET
CLEVELAND STR.
PALL MALL
COCKSPUR ST.
CHARING CROSS
STRAND
WHITE HALL
PARADE
ST JAMES'S PARK
PARLIAMENT STREET
BRIDGE STREET
WESTMINSTER BRIDGE
NEW PALACE YARD
QUEENS SQ.
PARK STR.
QUEEN STREET
THE BROAD WAY
TOTHILL STREET
Timber Wharf
Timber Yard
THE NEW ROAD

TOTHILL
FIELDS

RIVER
THAMES
WHITE CHAPEL
MILE END
WALWORTH COMMON

Rocque's large and detailed plan of the cities of Georgian London and the country ten miles round

41 **ROCQUE, John**

An exact Survey of the Cities of London and Westminster, the Borough of Southwark, The Country near ten Miles round; begun in 1741, finished in 1745, and publish'd in 1746, according to Act of Parliament, By John Rocque Land-Surveyor: Engrav'd by Richard Parr, and Printed by W. Pratt.

<u>Publication</u>
London, sold by the Proprietor John Rocque, next th Duke of Gaston's Head, in Hyde-Park Road, the Bottom of Picadilly, and at the Print Shops in London and Westminster, [1746].

<u>Description</u>
Large engraved wall map on 16 sheets, joined, and mounted on linen.

<u>Dimensions</u>
1450 by 1800mm (57 by 70.75 inches).

<u>Scale</u>
5½ inches to 1 statute mile.

<u>References</u>
Howgego 94 state (1).

One of the finest maps of – what is now – Greater London ever produced. It would appear that John Rocque, a French Huguenot, emigrated with the rest of his family to London in the 1730s, where he began to ply his trade as a surveyor of gentleman's estates, and with plans of Kensington Gardens, and Hampton Court. However, in 1737 he applied his surveying skills to a much great task, that of surveying the entire built-up area of London. Begun in the March of 1737, the map would take nine years to produce, eventually being engraved upon 24 sheets of copper and published in 1746. Whilst engaged upon this project Rocque was also working on the present map of the "Country ten miles round London", on a scale of 5½ inches to the mile, or one quarter of the scale of the large survey. The completed map was published in 1746.

An exact SURVEY of the
CITIES of *London* and *Weſtminſter,*
THE
BOROUGH of *Southwark,*
WITH
The *Country* near ten Miles round;
Begun in 1741, finiſhed in 1745, and publiſh'd in 1746, according to *Act of Parliament,*
By *JOHN ROCQUE* LAND-SURVEYOR:
Engrav'd by *RICHARD PARR*, and Printed by W. PRATT.
Sold by the Proprietor *John Rocque*, next the Duke of *Grafton's* Head, in Hyde-Park-Road, the Bottom of Piccadilly, and at the Print Shops in London and Weſtminſter, Price two Guineas in Sheets.

URBIUM MAXIME INSIGNIUM
LONDINI ET WESTMONASTERII,
NEC NON
MUNICIPII SUDOVERCENCIS,
Regionumque ad decem quaſi Milliaria cicumjacentium, Delineatio Ichnographica Anno 1741 *incobata,* 1745 *abſoluta*
A Johanne Rocque, *Topographo.*

CARTE TOPOGRAPHIQUE
Des Villes de *Londres, Weſtminſter,*
ET
Bourg de *Southwark,* et de leurs *Environs*
Levée trés exactement ſur les Lieux, en 1741, *achevée en* 1745, *et publiée ſelon un Acte de Parlement en* 1746.
Par *Jean Rocque.*

GROLES
GREEN
Temple Fortune
NORTH END
HAMPSTEAD HEATH
Ken Wood House
HAMPSTEAD
PONDS
Childs Hill Lane
The Cow House
The Blind Lane
Fortune Green
West End Lane
Shoot up hill Lane
HAMPSTEAD
Pound Street
Belsizebane
West End Lane
GREEN

GREENFORD GREEN
APPERTON
THE COURT FARM
HANGER HILL
EAST ACTON
ACTON COMMON
KENSINGTON
BROMPTON
HAMERSMITH
CHISWICK
FULHAM FIELDS
BARNES
BATTERSEA COMMON
RICHMOND PARK
PETERSHAM
HAM COMMON
PUTNEY HEATH
ROEHAMPTON
TWICKENHAM

UPPER HOLLOWAY
LOWER HOLLOWAY
NEWINGTON
NEWINGTON GREEN
SHAKLEWELL
KINGSLAND
OLDFORD
STRATFORD
WESTHAM
EASTHAM
PLASTOW LEWELLS

"Kings New Road"

42 **ROCQUE, John**

An exact Survey of the Cities of London and Westminster, the Borough of Southwark, The Country near ten Miles round; begun in 1741, finished in 1745, and publish'd in 1746, according to Act of Parliament, By John Rocque Land-Surveyor: Engrav'd by Richard Parr, and Printed by W. Pratt.

Publication
London, Sold by the Proprietor John Rocque, next the Barr, in Southampton-Street, Covent Garden. Printed by H. Lewis, at the Golden Ball, in Durham-Yard, [c.1751–4].

Description
Large engraved wall map on 16 sheets, joined, and mounted on linen.

Dimensions
1450 by 1800mm (57 by 70.75 inches).

Scale
5½ inches to 1 statute mile.

References
Howgego 94 state (5).

The present map is a fine example of the fifth state with minor alterations: for example the "Old Kings Road" is renamed the "Kings New Road", and "Peace Bridge", Barnes Common, to "Priests Bridge".

ADDINGTON
GREEN
The Pest House
Bayswatering
Marylebone
Turnpike
Tybourn
Tile Kiln
Cavendish Square
Royal Palace and Gardens
The Bason
H Y D E
P A R K
The Serpentine River
the Kings Old Road
the Kings New Road
Grosvenor Square
Brook Str.
Upper Brook Str.
Upper Grosvenor Str.
Mount Str.
Chappel Str.
South Str.
Hill Street
Hay Street
Charles Str.
Kensington
Gore
KNIGHTS BRIDGE
Hell House
Bell Lane
BROMPTON
Blackland Lane
Chelsea Common
Blackland School
THE FIVE FIELDS
Private Road
Bloody Bridge
Five Fields
QUEENS ELME
The Kings
Buckingham House
PIMLICO
Chelsea Road
Chelsea Water Works
CHELSEA
Hospital
Chelsea
The Worlds End
NEAT HOUSES
Battersea Wind Mills
FIELD

A reduction of Rocque's monumental map of Georgian London upon 24 sheets

43 **ROCQUE, John**

To Martin Folkes Esq[ui]r[e] President of the Royal Society: This Plan of the Cities of London.

Publication
London, Published according to Act of Parliament and Sold by the Proprietors John Tinney at the Golden Lion in Fleet Street, John Pine at the Golden Head in King Street near St. Ann's Church, Soho, and Thomas Bowles next the Chapter House in St, Paul's Church Yard. Where may likewise be had the Original Plan before mentioned, 20th May, 1749 .

Description
Engraved plan, title and text below, mounted on linen.

Dimensions
655 by 982mm (25.75 by 38.75 inches).

Scale
6 inches to 1 statute mile.

References
Howgego 100 (1).

The text below the plan states that the map was taken from Rocque's great survey of the city, the surveying of which was begun in March of 1737, and took nine years for Rocque to complete. The map is dedicated by John Pine and John Tinney to the president of the Royal Society, Martin Folkes. John Pine, who described himself as Bluemantel Pursuivant at Arms & Engraver of Seals, etc. to His Majesty, engraved the original 24-sheet map and the reduction. John Tinney was responsible along with Pine for the map's publication. The text below, which is dated 1742, states how accurate the 24-sheet map was and what lengths they went to: "they have not thought much of the trouble of drawing the main Plan over again before they began the engrave: and which last they have deferred, till we (ie Martin Folkes and P. Duvall) could venture to give them this additional Recommendation".

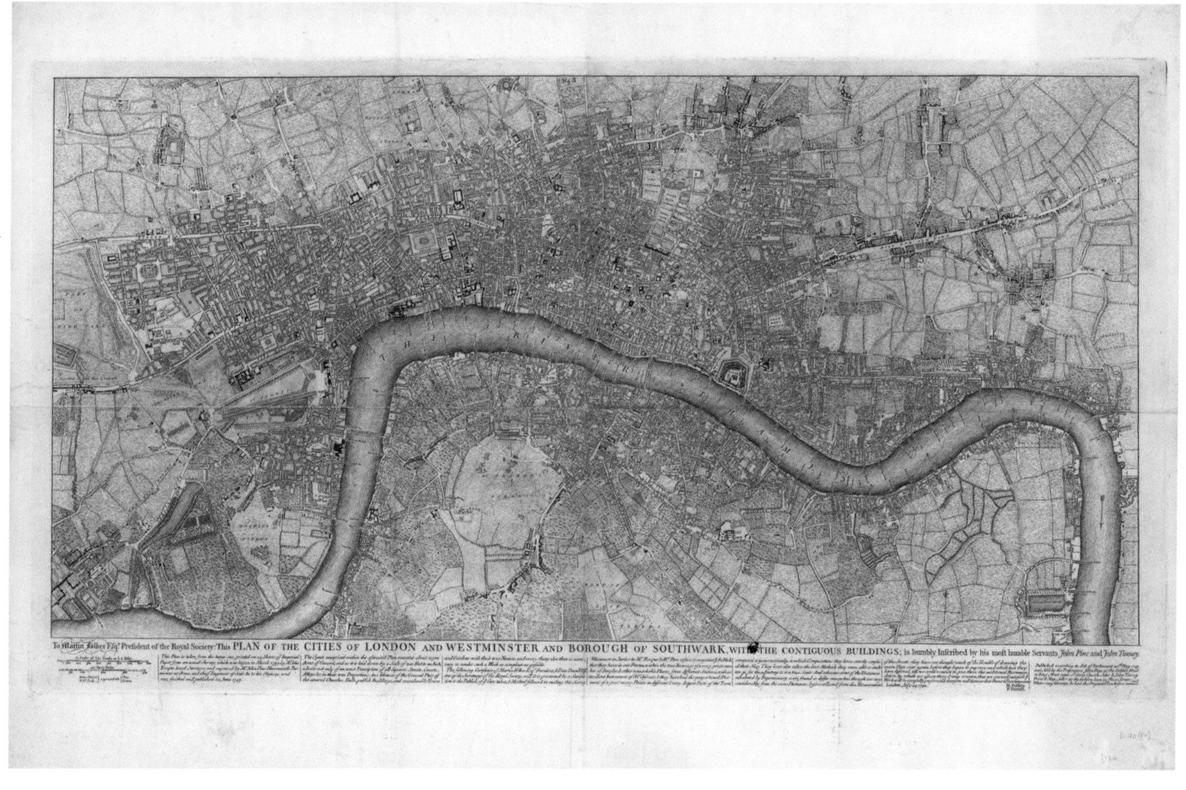

An unrecorded state with erasures on the plate

44 **ROCQUE, John**

To Martin Folkes Esq[ui]r[e] President of the Royal Society: This Plan of the Cities of London.

Publication
London, Published according to Act of Parliament, Sold by the Proprietors John Bowles in Cornhill, Carington Bowles in St. Pauls Church Yard, Robert Sayer in Fleet Street, and Thomas Jefferys at the Corner of St. Martins Lane in the Strand, 20th May, 1763.

Description
Engraved plan.

Dimensions
512 by 953mm (20.25 by 37.5 inches).

Scale
6 inches to 1 statute mile.

References
Howgego 100 (3).

The present map is an unrecorded state after Howgego 100 state (3). The imprint is the same as in state (3), however, the plan bears numerous erasures to the plate in order to accommodate new information: for example a large area of St Marylebone, Pimlico, Blackfriars Road, Blackfriars Bridge, New Bridge Street, and London Bridge.

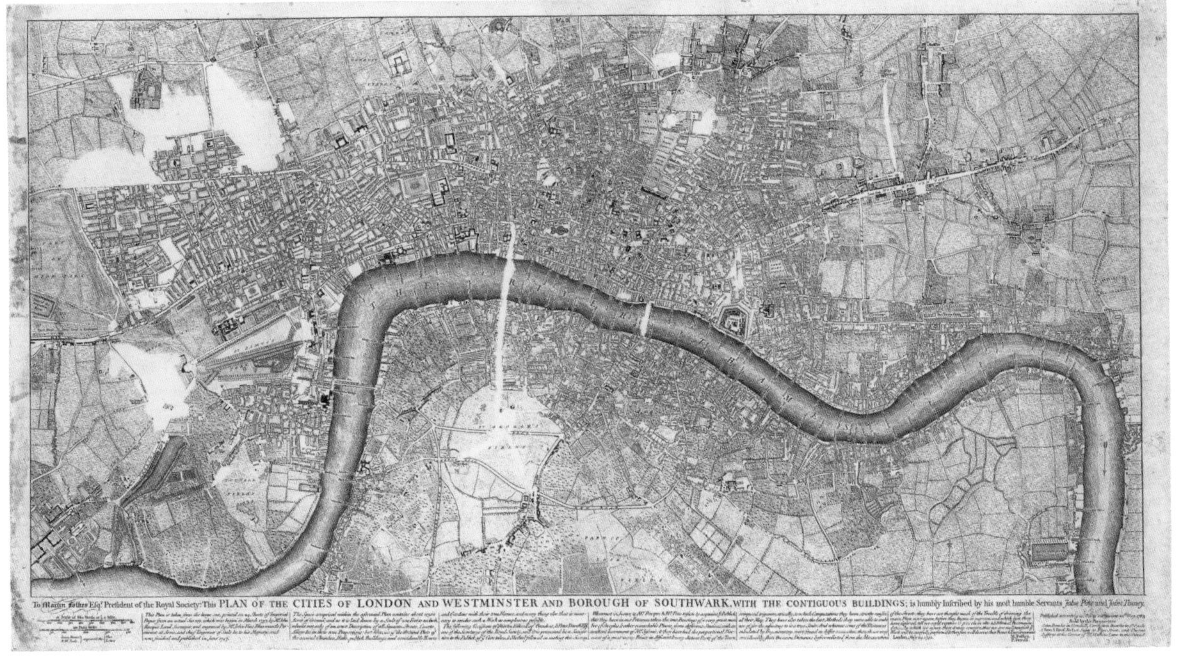

Magnificent complete set of six hand-coloured etched plans of the principal naval dockyards

45 **MILTON, Thomas**

[Complete series of Thomas Milton's Royal Dockyards] A Geometrical Plan and North Elevation of H.M. Dock Yard at Woolwich, with part of the Town... [&] A Geometrical Plan and West Elevation of H.M. Dock Yard and Garrison at Sheerness, with the Ordnance Wharfe... [&] A Geometrical Plan and West Elevation of His Majesty's Dock Yard near Plymouth, with the Ordnance Wharfe... [&] A Geometrical Plan and West Elevation of His Majesty's Dock Yard near Portsmouth, with part of the Common... [&] A Geometrical Plan, & North East Elevation Of His Majesty's Dock-Yard at Deptford, with Part of the Town... [&] A Geometrical Plan & North West Elevation of his Majesty's Dock-Yard at Chatham, with ye village of Brompton adjacent.

Publication
[London, T. Milton, 1753–1756].

Description
Six etched plans with views, engraved titles, key, coat of arms and dedication to cartouches within elaborate borders, conch shells suggested at corners, includes maritime vignettes of ships and ship building, also floral motifs, watermarked laid paper, 'Woolwich' with sheet trimmed, lacking lower margin, the others with wide margins, generally a little grubby and stained, with small closed tears, occasionally into plate.

Dimensions
675 by 490mm (26.5 by 19.25 inches).

Scale
6 inches to 1 statute mile.

References
Admiralty Library Manuscripts Portfolio B/44; DNB; See BL Maps K.Top.17.23.

An important set of plans by Thomas Milton (1743–1827), surveyor, draughtsman and engraver, a pupil of Woollett. He worked in Dublin and London and was governor of the Society of Engravers. The draughtsmanship of the shipping is attributed to John Cleveley, father of the marine artist of the same name, on Sheerness, Chatham and Plymouth. All six are rarely offered as a complete set as here. The British Library lists only four plans separately in their online catalogue.

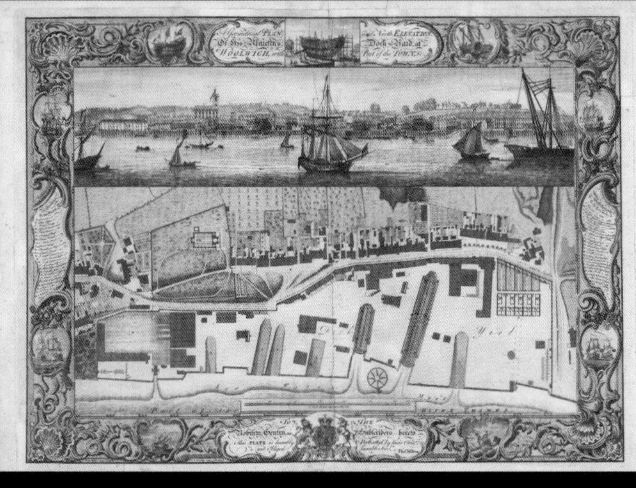

WOOLWICH
Dock Yard

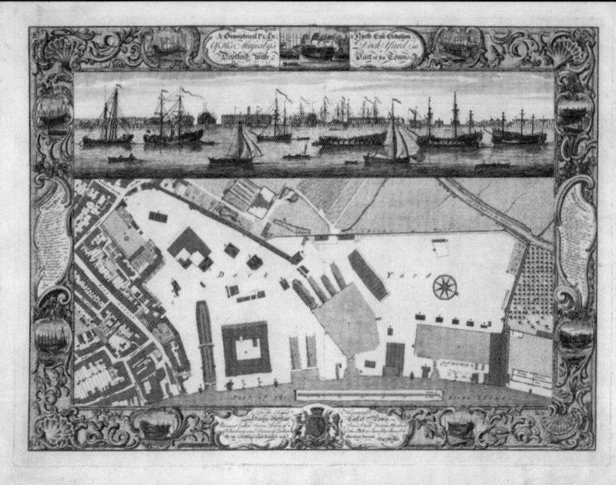

Dock Yard

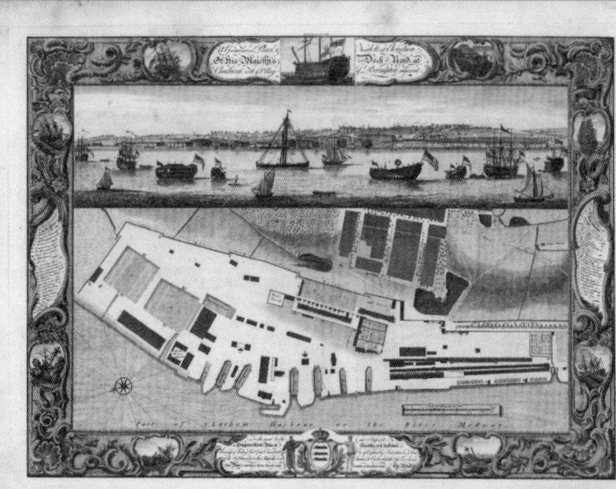

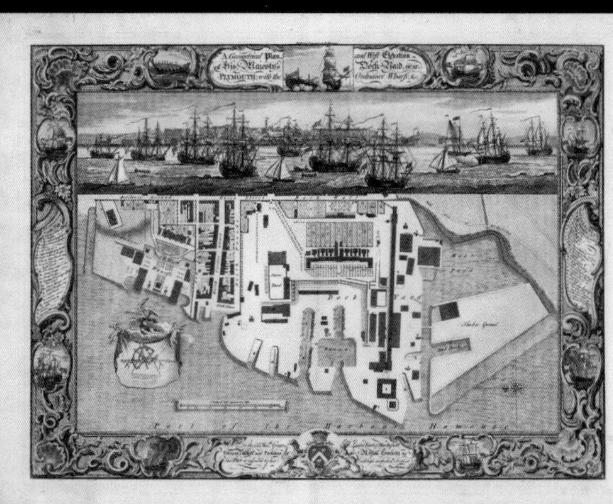

PLYMOUTH

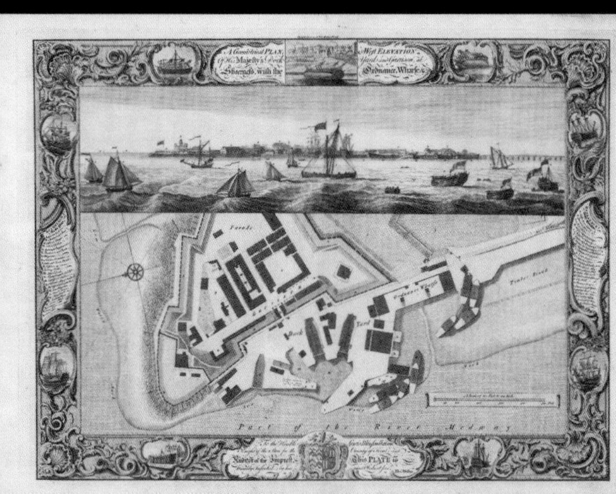

Part of the River Medway

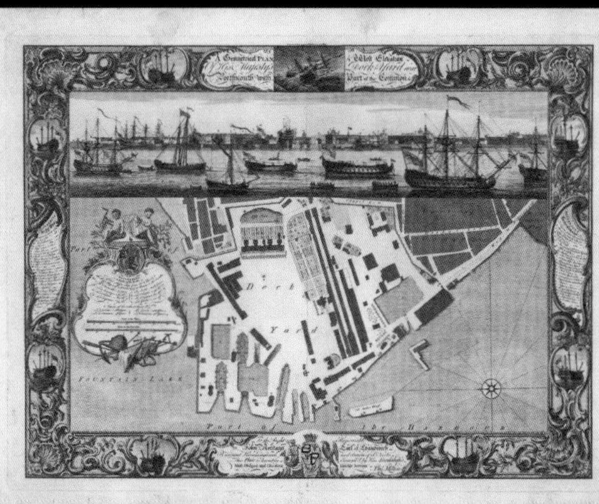

Dock
Yard

A Geometrical Plan,
of His Majesty's
PLYMOUTH; with the
Artillery Ground
Queen Street
Cannon Street
Ordnance
Canal
Wharfe
Gun Lane
North Corner Street
North Parade
North Corner
Store Yard
REFERENCES to the Plan
A. Gateway & Porters Lodge
B. Chapple
C. Commissioners House & Gard.
D. Officers Houses & Gardens
E. Offices
F. Store Houses
G. Differ. Workshops & Sheds
H. Engine & Pump Houses
I. Kilns to heat Plank
K. Double Dock
L. Single Dock
M. Launching slips
N. Boat Houses
O. Hemp Houses
P. Rigging House & Yarn houses
Q. Sail Loft
R. Rope Houses
S. Mast Houses
T. Tap House
V. Cranes
W. Magazines
X. Saw pits
Y. Stables
Z. Landing Places
A Scale of half quarter of a Mile
Part of the
To the Rt. Honble. George
Viscount Parker and President of
This Plate is inscrib'd by his
According to Act of Parliament

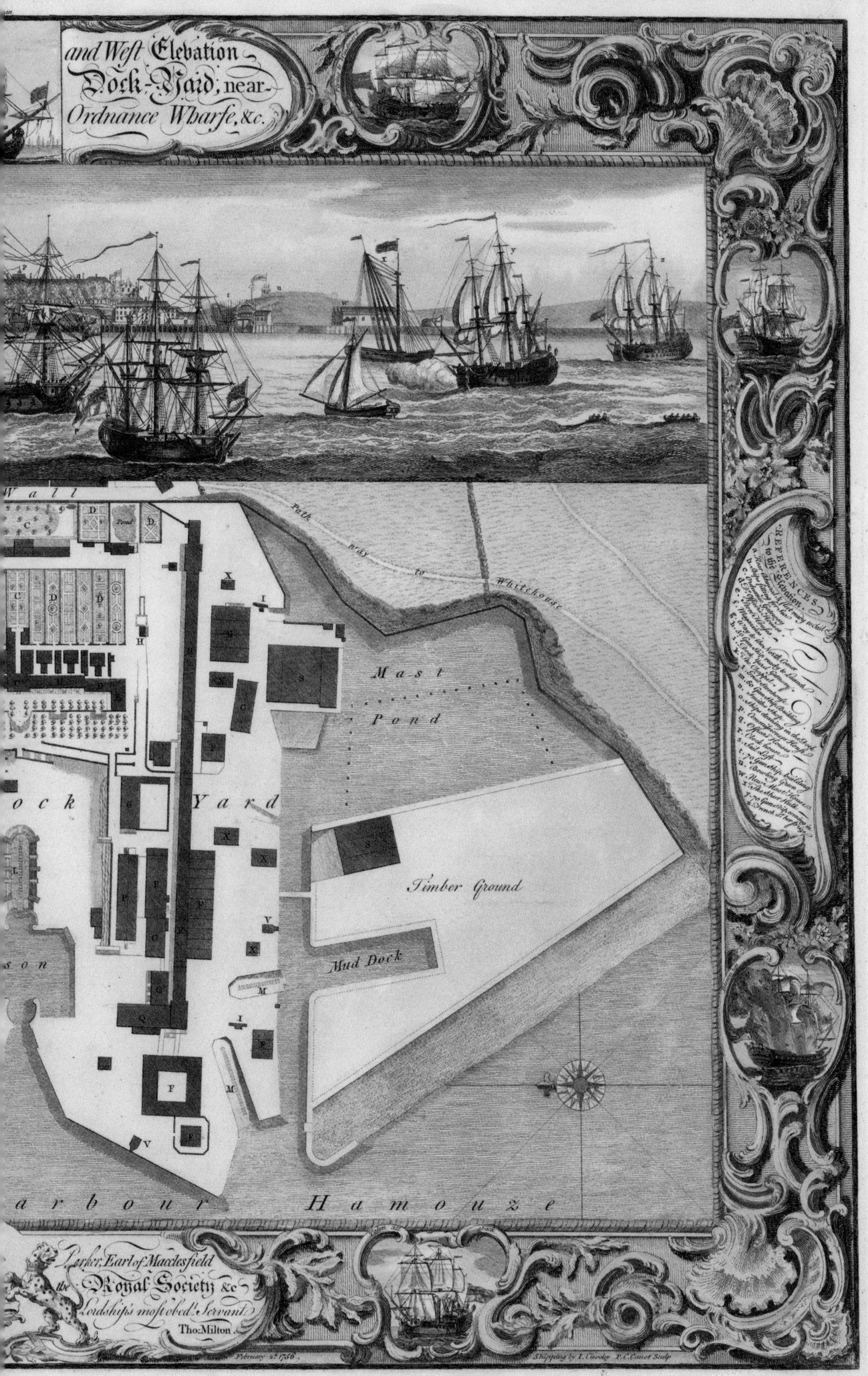
and West Elevation
Dock-Yard; near
Ordnance Wharfe, &c.
Wall
Path way to Whitehouse
Mast
Pond
ock
Yard
Timber Ground
Mud Dock
son
arbour
Hamouze
REFERENCES to the Elevation
Parker, Earl of Macclesfield
the Royal Society &c
Lordship's most obed.t Servant
Tho: Milton
February 2d 1756
Shipping by I. Cleveley P. C. Canot Sculp

Printed on silk

46 **ROCQUE, John**

A Plan of London on the same scale as that of Paris: In Order to ascertain the Difference of the Extent of these two Rivals, the Abbe de la Grive's plan of Paris, & that of London by J. Rocque have been divided into equal Squares where London Contains 39, and Paris but 29, so that the Superfice of London to that of Paris as 39 to 29 or as 5455 Acres to 4028. London therefore exceeds Paris, by 1427 Acres, the former being 8½ square Miles & Paris only 6⅓. by J. Rocque Chorographer to His Royal Highness the Prince of Wales, in the Strand, London 1754. R. Parr sculp. To the most High Puissant & Noble Prince John Duke of Montague &c. Grand Master of the most Honourable Order of the Bath, Master General of the Ordnance. Master of the Great Wardrobe, & Knight of the most Noble Order of the Garter, &c. &c. This Plan is most humbly Inscrib'd by his Graces most devoted & Obed.t humble Servant. John Rocque.

Publication
London, Publish'd according to Act of Parliament, 1754.

Description
Engraved map printed on silk, extending west to east from Hounslow to Woolwich, and north to south from Tottenham Hale to Croydon.

Dimensions
430 by 620mm (17 by 24.5 inches).

Scale
1¼ inches to 1 statute mile.

References
Howgego 101 state (1).

The map extends to what we now know as Greater London. It is based upon Rocque's map of 'London and The Country Near ten Miles Round' printed on 16 sheets and published in 1746. Above the plan the title appears in English and French, and states that the map is intended to be used to compare London and Paris "in order to ascertain the difference of the extent of these two rivals". The text goes on to say that London "exceeds Paris by 1427 Acres". Whether this deemed London the winner is unclear, as size is not everything. Below the plan is a dedication to the Duke of Montague, together with allegorical figures and the Duke's coat-of-arms. Rocque also produced a companion map of Paris in order for the gentleman to compare the cities side-by-side.

PART OF MIDDLESEX
PART OF SURRY
PART OF KENT
Prince JOHN Duke of MONTAGUE & Grand Master of
the most Honourable Order of the Bath, Master General of the Ordnance
Order of the Garter

The rare reduction on eight sheets of Rocque's monumental plan of London on 24 sheets.

47 **ROCQUE, John**

A Plan of the Cities of London and Westminster, the Borough of Southwark, and the Contiguous Buildings; with all the New Roads that have been made on account of Westminster Bridge, and the New Buildings and Alterations to the present year MDCCLV. Engraved from an Actual Survey made by John Rocque. This Plan extends from East to West near six Miles, and from North to South a little more than Three, and contains about 11,500 Acres of Ground. Published by the Proprietors of the Original Survey.

Publication
London, this plan was published according to Act of Parliament, 1755 [but c.1768].

Description
Engraved plan on eight sheets.

Dimensions
1150 by 1900mm (45.25 by 74.75 inches).

Scale
5½ inches to 1 statute mile.

References
Howgego 103 intermediate between state between (1) and (2).

Below the plan is a list of public buildings and churches. The advertisement states that the plan was based upon Rocque's 24 sheet plan, but was "improved by the Addition of all the new Roads made on account of Westminster Bridge, & all the Buildings and Alterations made since". The advertisement goes on to reiterate Martin Folkes's recommendation of the original plan's great accuracy. Martin Folkes was at the time President of the Royal Society and was a great advocate of Rocque's map. The text ends by mentioning both the 24 sheet and the single sheet reduction which may be had for the price of 5 shillings. Upon the plan are numerous erasures, most notably around Marylebone, Blackfriars Bridge, Blackfriars Road, New Bridge Street, and the London Wall. An approximate date can be ascertained due to the erasures: Blackfriars Bridge and Blackfriars Road were not open to the public until 1769.

The map is an intermediate state between Howgego 103 states (1) and (2). The erasures and change of imprint, (although not dated) give us an approximate date of 1768; therefore putting it after the first state. The imprint is again changed in state (2) and bears a date of 1775.

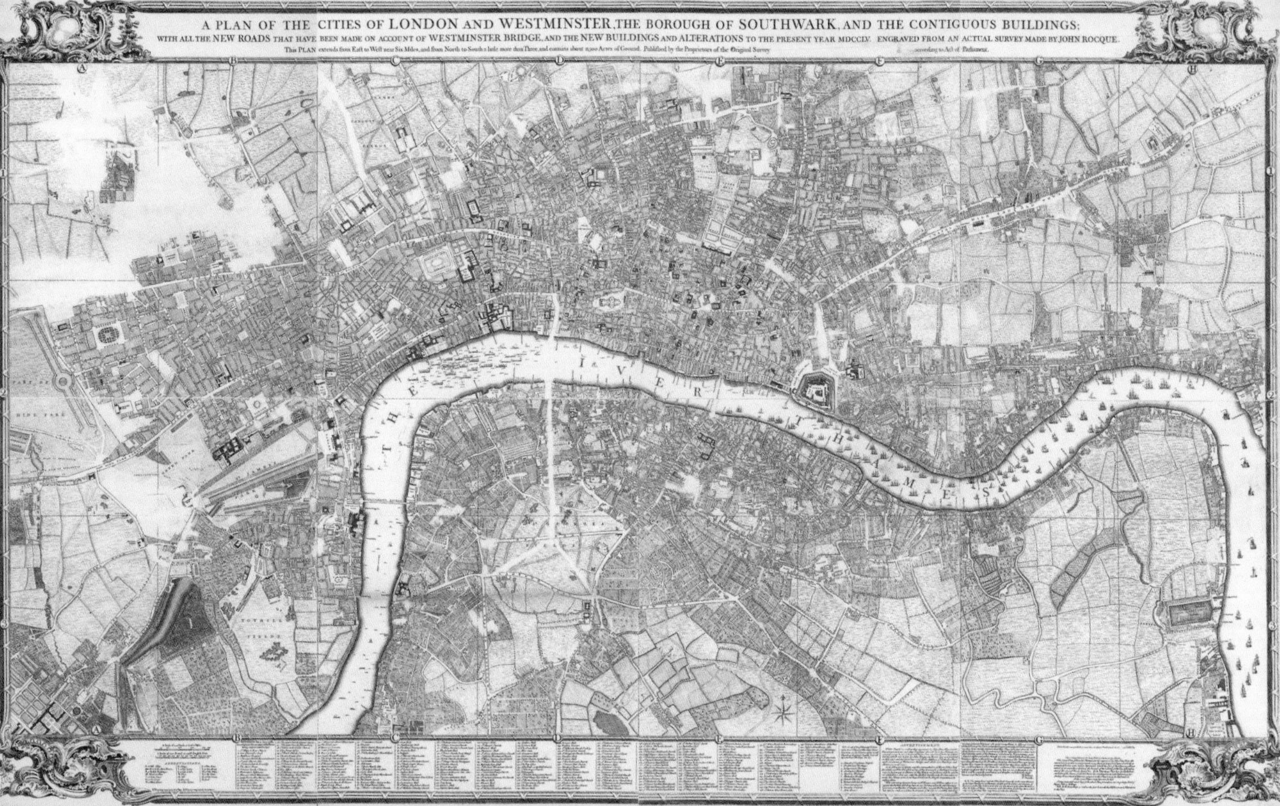
A PLAN OF THE CITIES OF LONDON AND WESTMINSTER, THE BOROUGH OF SOUTHWARK, AND THE CONTIGUOUS BUILDINGS;
WITH ALL THE NEW ROADS THAT HAVE BEEN MADE ON ACCOUNT OF WESTMINSTER BRIDGE, AND THE NEW BUILDINGS AND ALTERATIONS TO THE PRESENT YEAR MDCCLV. ENGRAVED FROM AN ACTUAL SURVEY MADE BY JOHN ROCQUE.
This PLAN extends from East to West near Six Miles, and from North to South a little more than Three, and contains about 11,500 Acres of Ground. Published by the Proprietors of the Original Survey
according to Act of Parliament.
THE RIVER THAMES
A
B
C
D
E
F
G
H

A fan of London

48 **BENNETT, Richard**

A New & Correct Plan of London, including all ye New Buildings &c.

Publication
[London, 1760].

Description
Engraved map, insets showing hackney coach fares, blank on verso, mounted as a fan on bone sticks with carved ivory end-pieces.

Dimensions
240mm diameter.

References
Not in Howgego; not in the Schreiber Collection.

A very rare map. The verso of the fan is stamped "Clarke & Co. No. 26, Strand, invent." Clarke was the inventor of a "new patent sliding fan".

"Few art forms combine functional, ceremonial and decorative uses as elegantly as the fan. Fewer still can match such diversity with a history stretching back 3000 years. Pictorial records show some of the earliest fans date back to around 3000BC, and there is evidence that the Greeks, Etruscans and Romans all used fans as cooling and ceremonial devices, while Chinese literary sources associate the fan with ancient mythical and historical characters. The first folding fans were inspired by and copied from prototypes brought into Europe by merchant traders and the religious orders that had set up colonies along the coasts of China and Japan. These early fans were regarded as a status symbol. While their "montures' (i.e. sticks and guards) were made from material such as ivory, mother of pearl, and tortoiseshell, often carved and pierced and ornamented with silver, gold and precious stones, the leaves were painted by craftsmen who gradually amalgamated into guilds such as The Worshipful Company of Fan Makers... The eighteenth century also saw the development of the printed fan: cheaper to manufacture and therefore cheaper to purchase, fans were suddenly available to a much wider audience than had previously been the case" (The Fan Museum).

Exhibited: "London: A Life in Maps", British Library, November 2006–March 2007.

A New & Correct PLAN of
LONDON,
including all ye New Buildings
&c.
Haberdashers Hospital
French Hospital
The Artillery Ground

Rare plan of St. George's Parish, Hanover Square

49 **BICKHAM, G[eorge]**

This Plate of St. Georges Parish, Hanover Square. With the Views of the Church and Chapels of Ease from the Original Survey of the late Mr Morris is most humbly Dedicated to the Right Hon.ble the Earl Lichfield & Sr Charles Tynte the Church Wardens and the rest of the Nobility and Gentlemen of the Vestry of the said Parish.

Publication
[London], Published according to Act of Parliament, & Sold in May's Buildings by G. Bickham Engraver, March the 24th, 1761.

Description
Broadsheet engraved plan, elaborate border with seven views of the parish's churches and chapels.

Dimensions
545 by 470mm (21.5 by 18.5 inches).

The plan bears a great deal of rococo decoration which not only encloses much of the map, but also the seven views of the parish churches and chapels of ease. Above the plan is a view of St. George's Hanover Square; to the left views of Berkeley Chapel, Audley Street Chapel, and Chelsea Chapel, and to the right Knightsbridge Chapel, Conduit Street Chapel, and Mayfair Chapel. Below is a key together with a scale-bar, the imprint, and a dedication reading.

St. George's Church was built in 1721–4 to the designs of John James and at a cost of £10,000, as one of fifty churches projected by Queen Anne's Act of 1711. The new parish was carved out of that of St. Martin-in-the-Fields and covered what are now the areas of Mayfair, Belgravia and Pimlico. The church itself was the site of many high-profile weddings, including those of the future U.S. President Theodore Roosevelt and the architect John Nash. George Frederick Handel was a frequent worshipper at St. George's, which is now home to the annual Handel Festival.

Audley Street Chapel, now Grosvenor Chapel, was built in 1730–31 by Benjamin Timbrell, and was where American servicemen worshipped during World War II. It is mentioned in John Summerson's Georgian London: "The chapel-of-ease in South Audley Street is, externally, an unattractive building in brown brick and stone, though the interior has merit… It is very much a craftsman's church… it is more like some of the churches of Maine, Connecticut and Massachusetts than the other London churches of the time."

Berkeley Chapel, which dates from around 1750, was pulled down in 1907 and replaced with Charles House. Sydney Smith, the noted Anglican cleric, often preached there in the early nineteenth century, and his success there was such that there was often not even room for standing.

Knightsbridge Chapel began as a chapel on Knightsbridge Green attached to a leper hospital founded by Westminster Abbey. The chapel was rebuilt in 1629, further rebuilt in 1699, and renovated again in 1789. A new church, Holy Trinity Knightsbridge, was erected on its site in 1861. In 1901 it was demolished and rebuilt on a new site in Prince Consort Road, where it is still open.

Conduit Street (or Trinity) Chapel, was built on the site of a wooden field chapel built by James II. It was conveyed on wheels to wherever the king travelled, to be used for his private masses. In 1686 it was in his camp at Hounslow Heath, where it stayed until some time after the Revolution. It was then removed to Curzon Street, where it was later rebuilt in more durable materials and used a chapel by the neighbouring inhabitants. The chapel was demolished in 1875.

Mayfair (later named Curzon) Chapel was first erected in 1730, and was the location of numerous clandestine marriages, before the Marriage Act of 1753, including those of the Duke of Chandos and Mrs. Anne Jeffrey in 1744. The chapel was pulled down in 1899 and replaced by Sunderland House, built for the 9th Duchess of Marlborough.

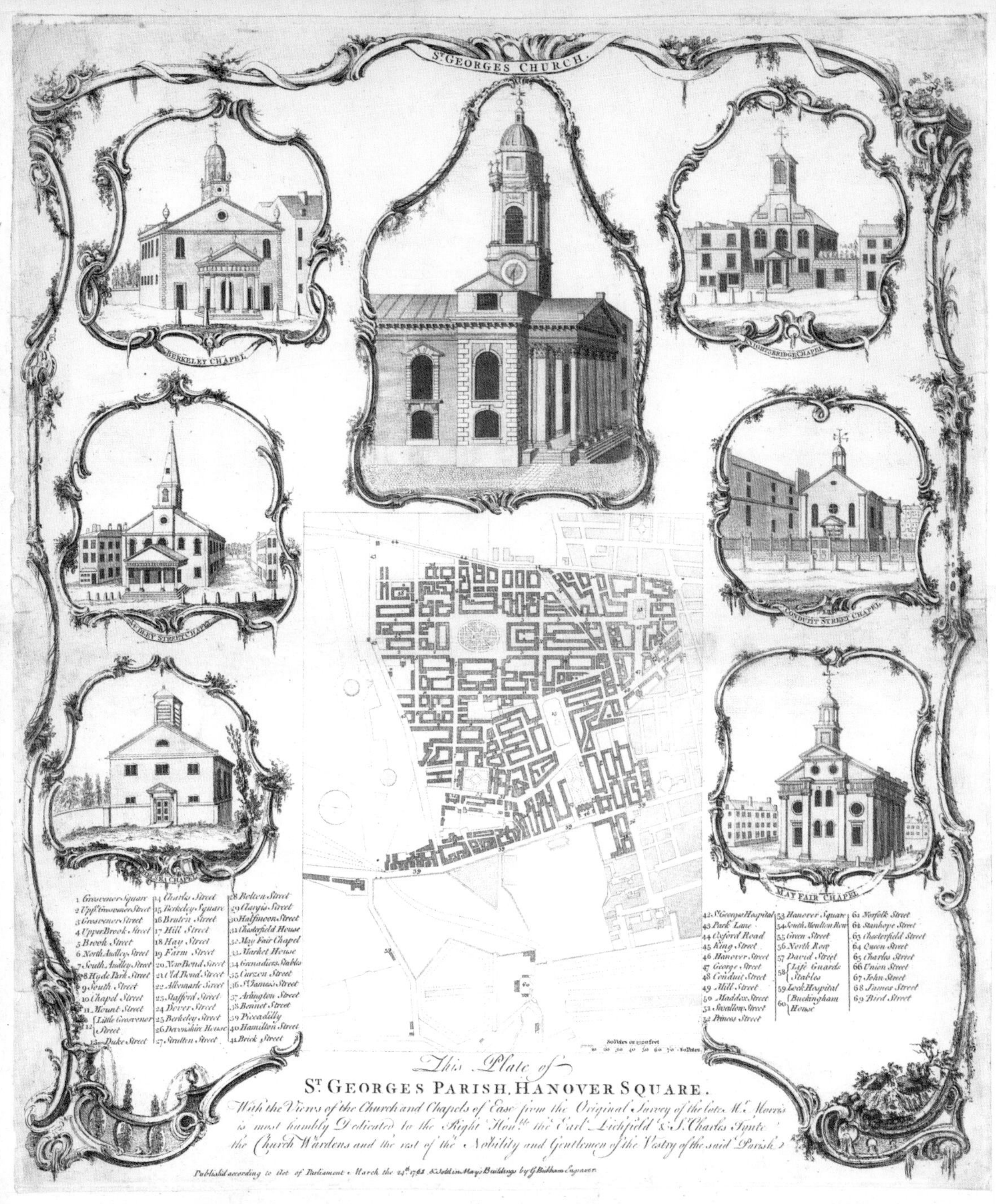

St. GEORGES CHURCH.
BERKELEY CHAPEL
KNIGHTSBRIDGE CHAPEL
AUDLEY STREET CHAPEL
CONDUIT STREET CHAPEL
MAY FAIR CHAPEL
1 Grosvenor Square
2 Upp. Grosvenor Street
3 Grosvenor Street
4 Upper Brook Street
5 Brook Street
6 North Audley Street
7 South Audley Street
8 Hyde Park Street
9 South Street
10 Chapel Street
11 Mount Street
12 Little Grosvenor Street
13 Duke Street
14 Charles Street
15 Berkeley Square
16 Bruton Street
17 Hill Street
18 Hay Street
19 Farm Street
20 New Bond Street
21 Old Bond Street
22 Albemarle Street
23 Stafford Street
24 Dover Street
25 Berkeley Street
26 Devonshire House
27 Strutton Street
28 Bolton Street
29 Clargis Street
30 Halfmoon Street
31 Chesterfield House
32 May Fair Chapel
33 Market House
34 Grenadiers Stables
35 Curzon Street
36 St. James's Street
37 Arlington Street
38 Bennet Street
39 Piccadilly
40 Hamilton Street
41 Brick Street
42 St. Georges Hospital
43 Park Lane
44 Oxford Road
45 King Street
46 Hanover Street
47 George Street
48 Conduit Street
49 Mill Street
50 Maddox Street
51 Swallow Street
52 Princes Street
53 Hanover Square
54 South Moulton Row
55 Green Street
56 North Row
57 David Street
58 Life Guards Stables
59 Lock Hospital
60 Buckingham House
61 Norfolk Street
62 Stanhope Street
63 Chesterfield Street
64 Queen Street
65 Charles Street
66 Union Street
67 John Street
68 James Street
69 Bird Street
80 Poles or 1320 feet
10 20 30 40 50 60 70 80 Poles
This Plate of
St. GEORGES PARISH, HANOVER SQUARE.
With the Views of the Church and Chapels of Ease from the Original Survey of the late Mr. Morris
is most humbly Dedicated to the Right Honble. the Earl Lichfield & Sr. Charles Tynte
the Church Wardens and the rest of the Nobility and Gentlemen of the Vestry of the said Parish
Publish'd according to Act of Parliament March the 24th. 1762 & Sold in May's Buildings by G. Bickham Engraver.

Fishy business at Westminster…

50 **FOURDRINIER, C[harles]**

A Plan of Part of the Ancient City of Westminster from College Street to Whitehall, and from the Thames to St. James's Park, in which are laid down all the New Streets that have been built & other alterations made since the Building of Westminster Bridge.

Publication
[London], Published according to act of Parliament by C. Fourdrinier & Co. at Charing Cross, Jan. 1761.

Description
Separately issued engraved plan, key to old and new buildings, a few nicks and tears skilfully repaired.

Dimensions
550 by 770mm (21.75 by 30.25 inches).

The plan shows Whitehall undergoing a significant amount of change, with new buildings marked in dark grey and old in light grey. Much of Parliament Street, Great George Street, and Abingdon Street is new. Along Whitehall, Downing Street is clearly marked with the new Treasury building facing on to Horse Guards Parade. Next to the Treasury is a building marked 'Sir Matthew Featherstones'. Now known as Dover House and home to the Scotland Office, it was originally designed by James Pine in the 1750s for Sir Matthew Fetherstonhaugh. Westminster Abbey boasts its Neo-Gothic West Towers, constructed by Nicholas Hawksmoor between 1722 and 1745.

Westminster Bridge, noted as "The New Bridge" on this map, had recently been completed. After years of opposition, the construction of the bridge finally received Parliamentary approval in 1736 and was built between 1739–1750. Just next to the Bridge can be seen a fish market and Fish Market street. This was a relevantly recent addition, as in 1749, an act was passed "for making a free Market for the sale of fish in Westminster", as the inhabitants of Westminster had "long laboured under the want of a fish-market, and complained that the price of this species of provision was kept up at an exorbitant rate by the fraudulent combination of a few dealers, who engrossed the whole market of Billingsgate…" (The History of England, vol. 6, p.65).

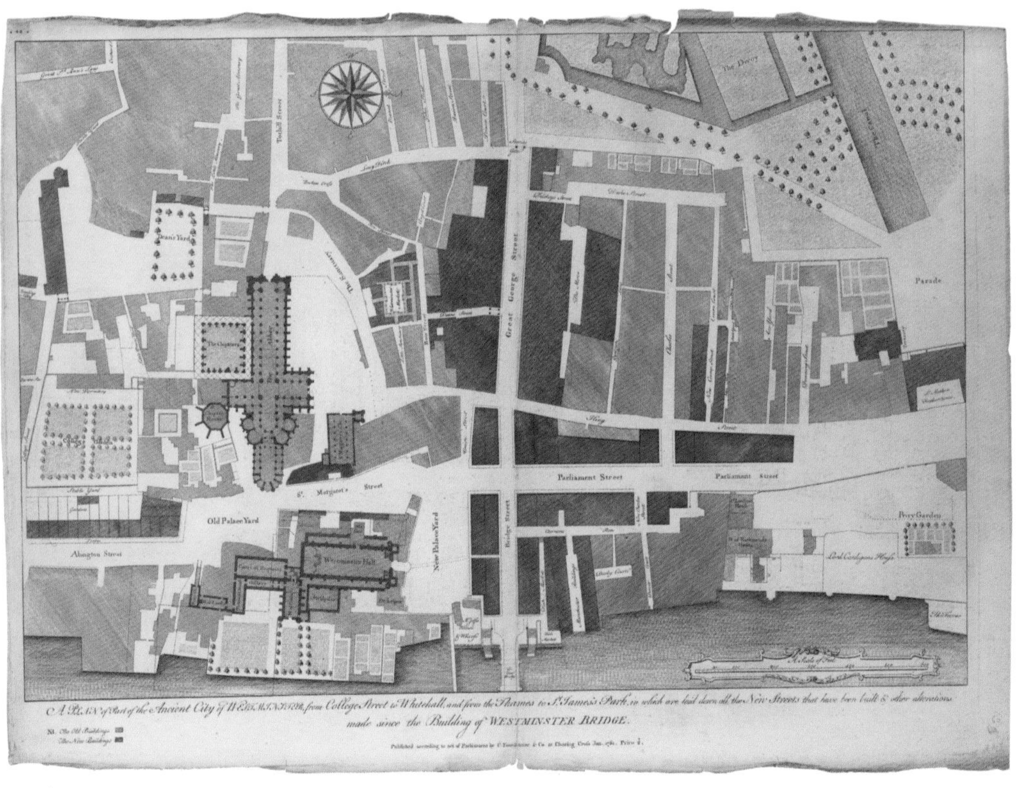

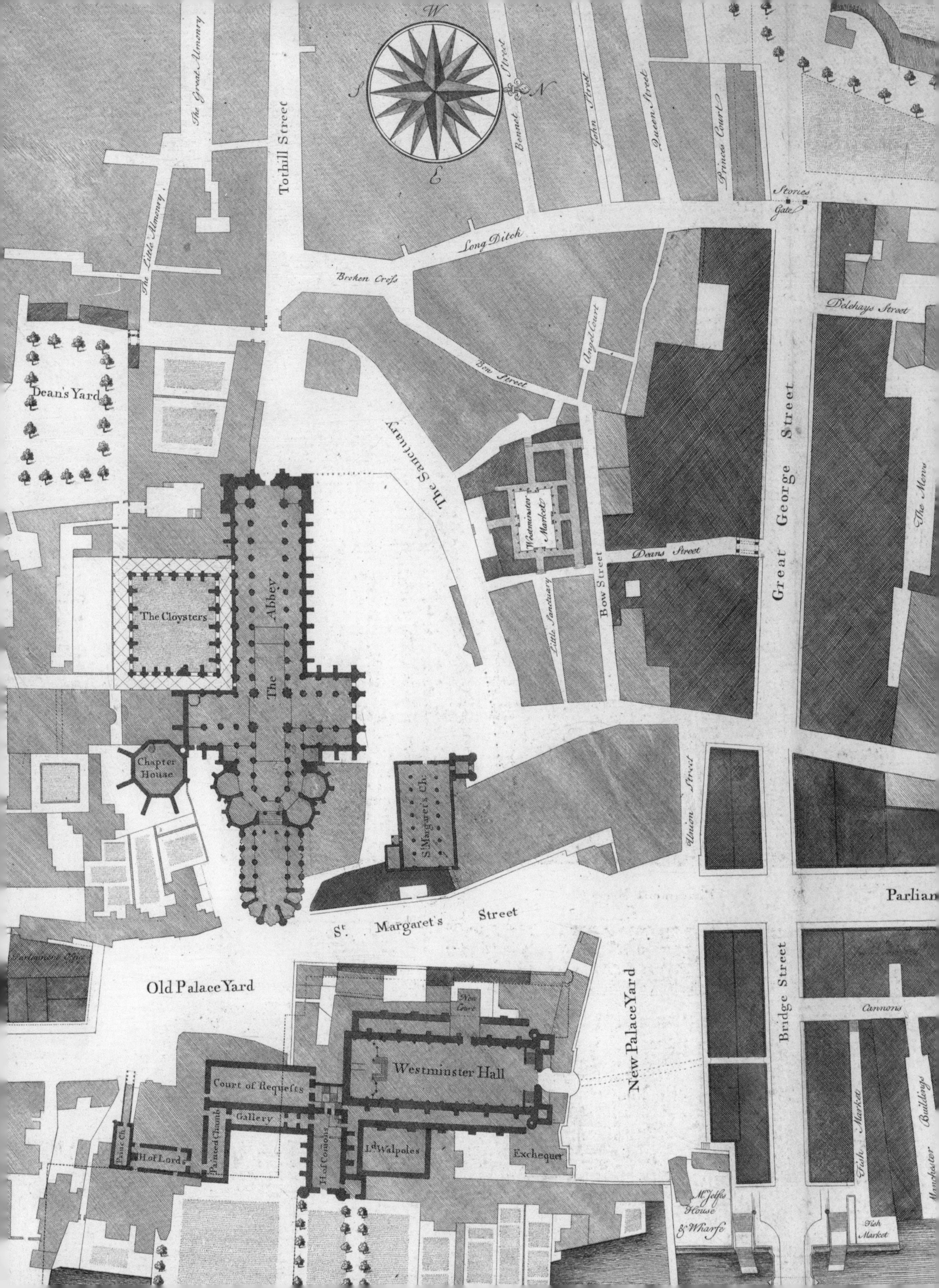
W
S
N
E
The Great Almonry
Tothill Street
Bennet Street
John Street
Queen Street
Princes Court
Stories Gate
The Little Almonry
Long Ditch
Broken Cross
Delahays Street
Angel Court
Bow Street
Dean's Yard
The Sanctuary
Westminster Market
Deans Street
Great George Street
The Mews
The Abbey
The Cloysters
Little Sanctuary
Bow Street
Chapter House
St. Margarets Ch.
Union Street
Parliam
St. Margaret's Street
Old Palace Yard
New Palace Yard
Bridge Street
New Court
Cannons
Westminster Hall
Court of Requests
Gallery
Painted Chamb.
H. of Lords
Ld. Walpoles
Exchequer
H. of Comons
Fish Market
Manchester Buildings
Mr. Jelfs House & Wharfe
Fish Market

Rare four sheet reduction of Rocque's map of Greater London

51 **ROCQUE, Mary Ann**

The Environs of London Reduced from an Actual Survey in 16 Sheets, by the Late John Rocque, topographer to His Majesty with New Improvements to the Year 1763. To the Right Honorable George Montague Earl of Cardigan, Baron Brudenell &c. This Plan of the Environs of London Is humbly dedicated By His Lordship's most humble & obliged servant, Mary Ann Rocque.

Publication
London, Printed for Carington Bowles, No. 69 in St. Paul's Church Yard, & Robert Sayer, No. 53 in Fleet Street, 1763.

Description
Engraved map, on four sheets, splits to centrefolds skilfully repaired.

Dimensions
900 by 1300mm (35.5 by 51.25 inches).

Scale
2½ inches to 1 statute mile.

References
Howgego 124 (1a).

This fine survey is a reduction of John Rocque's monumental survey of London upon 16 sheets. The map bears the name of Rocque's widow, Mary Ann, who continued his map publishing business. The plan also bears the imprint of Carington Bowles; to whom Mary would transfer much of the business at some time around 1769.

THE
ENVIRONS of LONDON
in 16 Sheets, by the Late
Topographer to His
MAJESTY
PART OF ESSEX
PART OF SURRY
PART OF KENT

Dicey's famous "elelixir salutis"

52 **DICEY, Cluer**

A New & Accurate Plan of the Cities of London and Westminster & Borough of Southwark

Publication
London, Printed according to Act of Parliament and Sold by C. Dicey and Co. in Aldermary Church Yard, 1765.

Description
Engraved map on two sheets, hand-coloured in outline, inset views of St James's Palace, Monument, St. Paul's Cathedral, a view of the Cities of London and Westminster, Westminster Abbey, and Buckingham Palace (named Queens Palace), minor paper restoration to margins not affecting image.

Dimensions
600 by 1050mm (23.5 by 41.25 inches).

Scale
c. 7½ inches to one statute mile.

References
Howgego 133 (1).

There are several vignette views on the plan, most notably the view of the Cities of London and Westminster looking upstream, and a view of Buckingham Palace – named the Queens Palace. The Palace was often called such throughout the eighteenth century due to it being the private residence of Queen Charlotte, the wife of George III; the majority of state business was carried out at St James's Palace. On the plan the City is marked out by stippling, with the parishes and wards marked by roman numerals; many of the most important buildings are shown pictorially, and the Thames is packed with ferrymen, barges, and sailing vessels. To the lower left is a table of watermen and hackney carriage rates, and a list of the 26 wards of the City of London.

The Diceys were well known as publishers of 'chapbooks', cheap pocket-sized pamphlets which were circulated through a network of chapmen – hawkers and pedlars who would travel around England, attending markets, fairs and so on, offering the books, ballads, portraits, maps and topographical prints.

Interestingly, they did not restrict themselves to the printing business, but were also involved in patent medicine, becoming one of the largest distributors of "Daffy's original and famous elixir salutis", advertised as "the choice drink of health, or, health-bringing drink. Being a famous cordial drink, found out by the providence of the Almighty, and (for above wenty [sic] years,) experienced by my self and divers persons (whose names are at most of their desires here inserted) a most excellent preservative of man-kind. A secret far beyond any medicament yet known, and is found so agreeable to nature, that it effects all its operations, as nature would have it…"

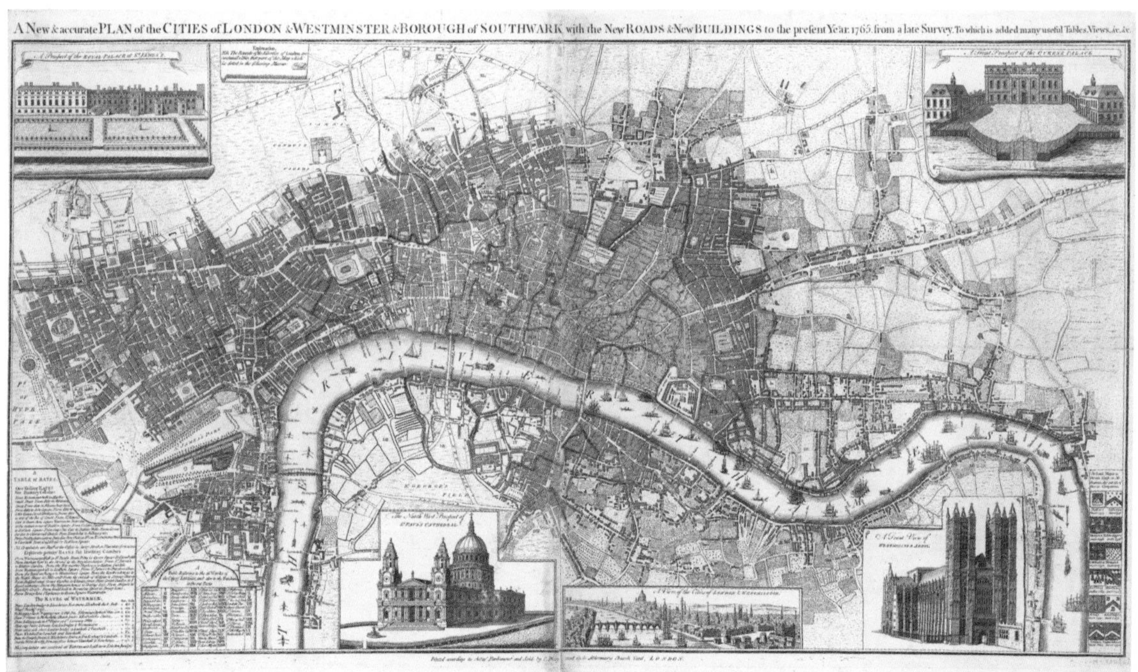

53 **DICEY, Cluer**

A New & Accurate Plan of the Cities of London & Westminster & Borough of Southwark

Publication
London, Printed according to Act of Parliament and Sold by I. Marshall and Co. in Aldermary Church Yard, 1789.

Description
Engraved map on two sheets, joined, hand-coloured in outline, minor paper restoration to margins not affecting image.

Dimensions
700 by 1120mm (27.5 by 44 inches).

Scale
c. 7½ inches to one statute mile.

References
Howgego 133 (6).

The plan was first published by Cluer Dicey in 1765. By the time the present map was printed, Dicey had passed away and the plates had come into the hands of John Marshall, with whom Dicey had had a long and fruitful partnership.

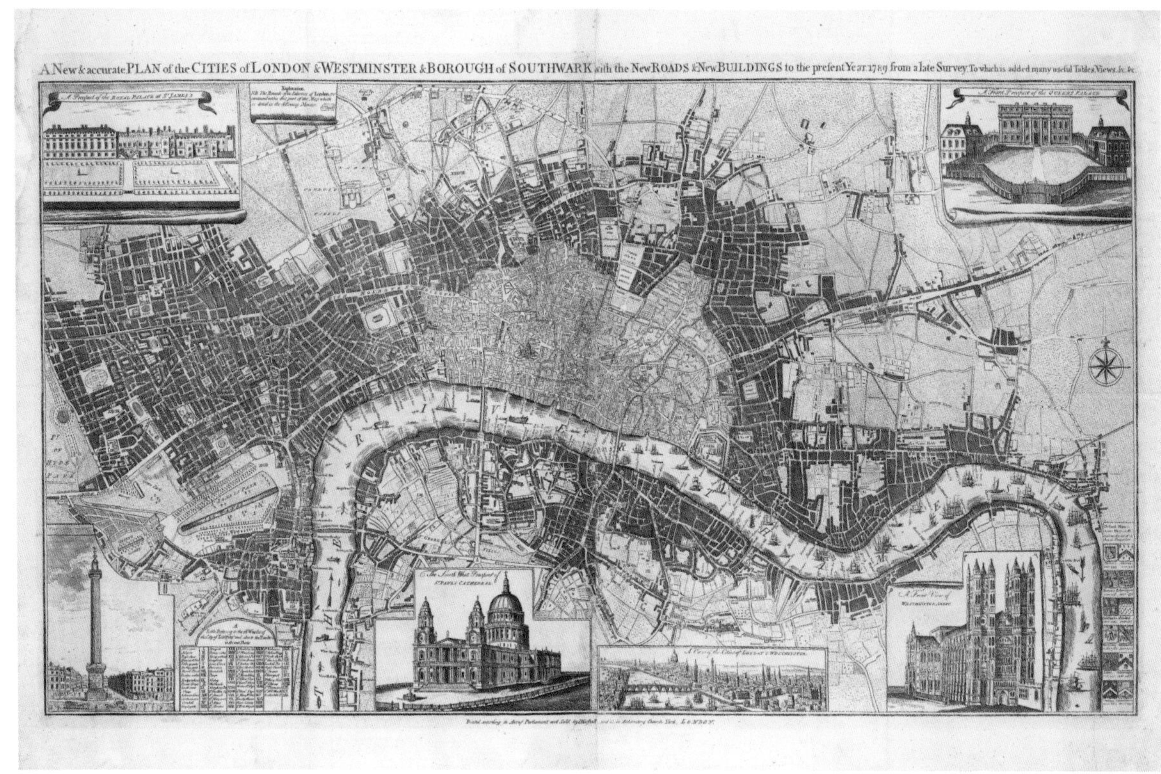

Rare and detailed plan of St James's

54 **RHODES, William**

A Correct Plan of the Parish of St. James's Westminster 1770 by William Rhodes.

Publication
[London], William Faden, Charing Cross, 1770.

Description
Broadsheet engraved plan, hand-coloured in outline, a few small tears to margins, skilfully repaired.

Dimensions
500 by 400mm (19.75 by 15.75 inches).

The map was drawn by the surveyor, William Rhodes of Great Marlborough, in 1770 at the request of a committee of the House of Commons, which was considering amendments to a paving act (Westminster Library, D1764, 27 April 1776). In 1777 Rhodes sold his survey to the parish for £50 and in the following year a mere one hundred copies were printed from an engraving by William Faden.

The plan stretches west to east from Old Bond Street to Wardour Street, and from north to south from Oxford Street to Pall Mall. It highlights several important buildings, including Spencer House, Burlington House and Gardens, St James's Church on Piccadilly, the 'Opera House' on Haymarket, now the site of 'Her Majesty's Theatre', several chapels, an auction room, and St. James's School, on Boyle Street, one of the oldest schools in London.

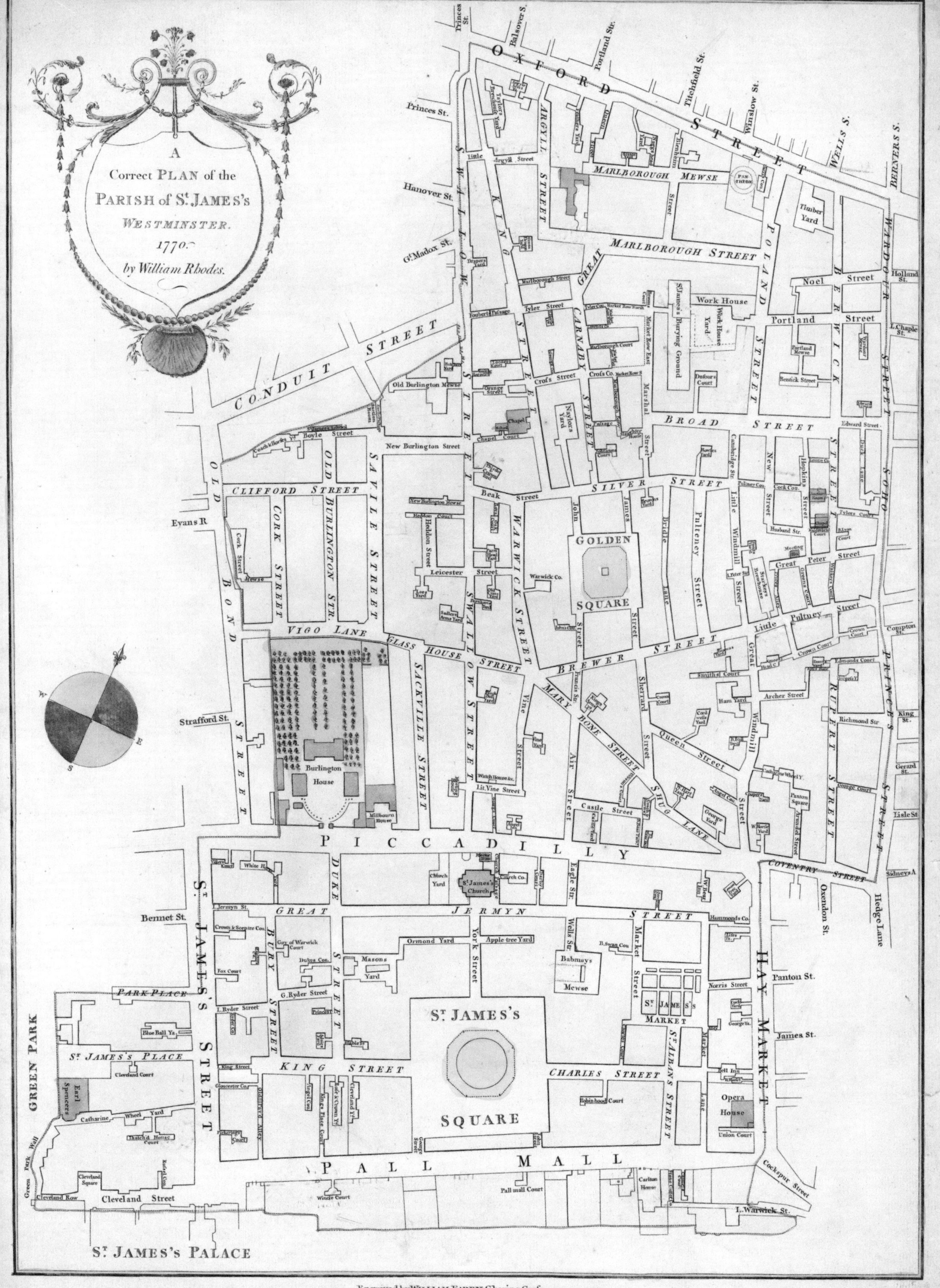
A Correct PLAN of the PARISH of St. JAMES's WESTMINSTER. 1770.
by William Rhodes.
OXFORD STREET
CONDUIT STREET
PICCADILLY
PALL MALL
GOLDEN SQUARE
St. JAMES's SQUARE
St. JAMES's STREET
St. JAMES's PALACE
GREEN PARK
HAY MARKET
Burlington House
St. James's Church
Opera House
Engraved by WILLIAM FADEN Charing Cross.

The Great Offices of State and The Savoy

55 **BOWLES, Carington**

The Traveller's Guide through London, Westminster, and Borough of Southwark; with their Liberties: Exhibiting the Streets, Roads, Churches, Palaces, Public Buildings, &c., as they have been lately extended and improved, By Act of Parliament.

Publication
London, Printed for Carington Bowles, Map & Printseller, at No. 69 in St. Pauls Church Yard, 31st Augt. 1770.

Description
Engraved plan on three sheets, joined, original hand-colour in outline, trimmed to neatline with minor loss to old folds skilfully repaired.

Dimensions
450 by 900mm (17.75 by 35.5 inches).

Scale
c.6½ inches to one statute mile.

References
Howgego 147a (1).

The plan has numerous tables, which contain information on the Great Offices of State, such as the master of the horse and the Board of Green Cloth; the Chief Magistrates of Westminster; and most interestingly the Parishes in Surrey and Middlesex which were in and outside the boundaries of the Bills of Mortality. The Bills of Mortality were set up at the end of the sixteenth century, in response to an outbreak of the plague. They initially recorded the number of deaths upon a weekly basis, however, by the time of the publication of the map the bills included baptisms, cause of death and age of the deceased. The plan also shows the newly built Westminster and Blackfriars Bridges, together with the new roads. The Savoy is marked (later the site of the hotel) as belonging to the Duchy of Lancaster. The Duchy then, as now, is used to provide income for the British monarch.

The plan bears a striking similarity to Thomas Jeffery's 'A New Plan of the City and Liberty of Westminster…' (item 56).

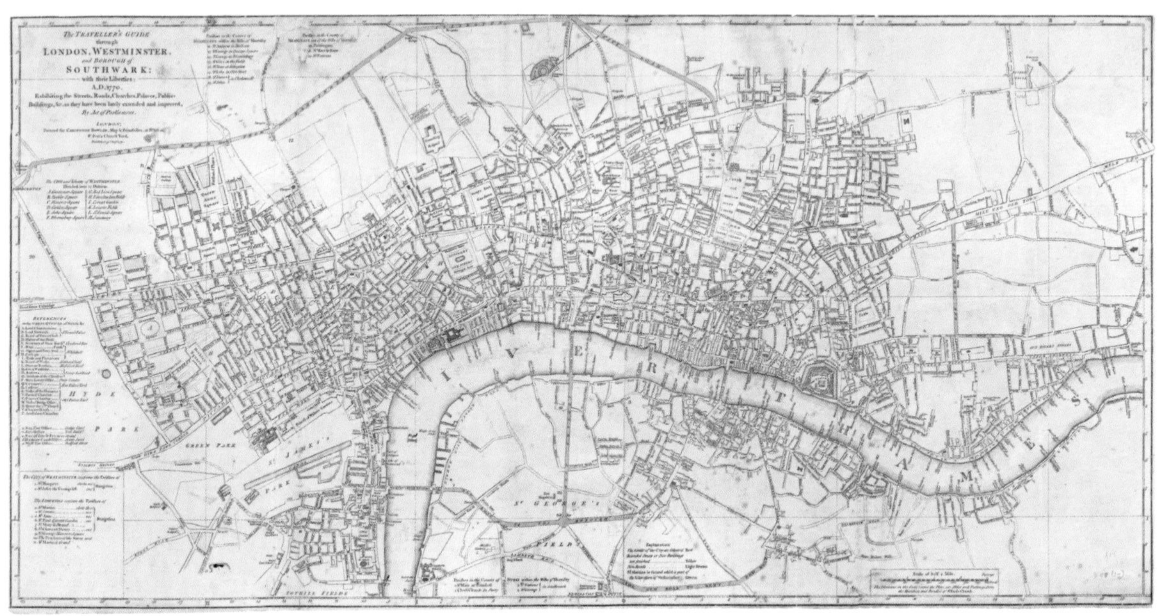

Information "Not extant on any other Plan"

56 **JEFFERYS, Thomas**

A New Plan of the City and Liberty of Westminster... [and] A New Plan of the city of London and Borough of Southwark, Exhibiting all the New Streets, Roads &c. Not extant in any other Plan.

Publication
[London, Thomas Jefferys, c. 1772].

Description
Two engraved companion maps, one of Westminster, and the other of the City of London and Southwark, original hand-colour in outline.

Dimensions
470 by 585mm (18.5 by 23 inches).

Scale
6¾ inches to one statute mile.

References
Howgego 122 (3).

These two companion plans exhibit a great deal of detailed information. Jefferys' boast that much of the information was "Not extant on any other Plan" is in many ways borne out by the numerous tables, which contain information on the Great Offices of State, such as the master of the horse and the Board of Green Cloth; the Chief Magistrates of Westminster; and, most interestingly, the Parishes in Surrey and Middlesex which were in and outside the boundaries of the Bills of Mortality. The Bills of Mortality were set up at the end of the sixteenth century, in response to an outbreak of the plague. They initially recorded the number of deaths upon a weekly basis, however by the time of the publication of the map the bills included baptisms, cause of death and age of the deceased. The plan also shows the newly built Westminster and Blackfriars Bridges, together with the new roads. The Savoy is marked (later the site of the hotel) as belonging to the Duchy of Lancaster. The Duchy then, as now, is used to provide income for the British monarch.

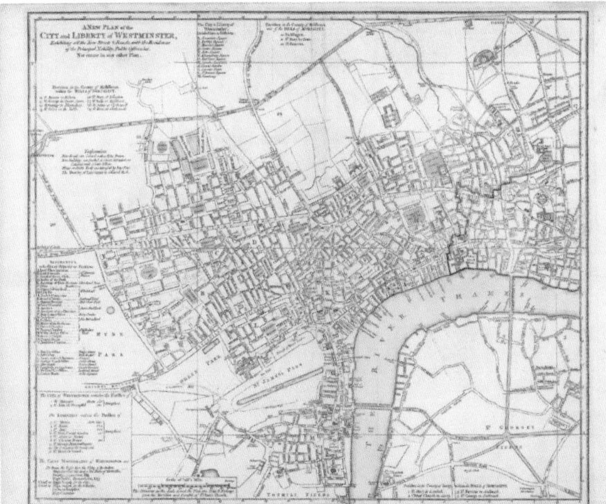

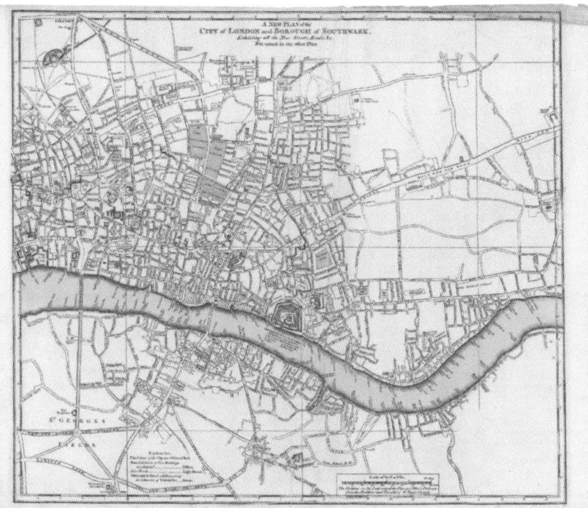

“Upwards of 350 Hackney coach Fares”

57 **CARY, James**

London, Westminster and Southwark...

Publication
London, Published as the Act directs by J. Wallis, No. 16 Ludgate Street, and by [damaged] Strand. Jan. 1st, 1786.

Description
Engraved plan, dissected and mounted on linen, trimmed to neatline.

Dimensions
420 by 552mm (16.5 by 21.75 inches).

Scale
4 inches to 1 statute mile.

References
Howgego 173 (5).

As stated in the title the map gives a list of: “Upwards of 350 Hackney coach Fares laid down by actual Measurement, and the prices regulated agreeable to the late act of Parliament 1786”. By the Act, the coachmen were allowed to charge by either time or distance: for 1s you could travel a mile and a quarter, or spend three-quarters of an hour in a cab. Anyone found exceeding the rates laid down were liable to be fined and summoned before the Hackney Coach Office at Somerset Place. The plan also gives the “rates of oars” up and down the Thames.

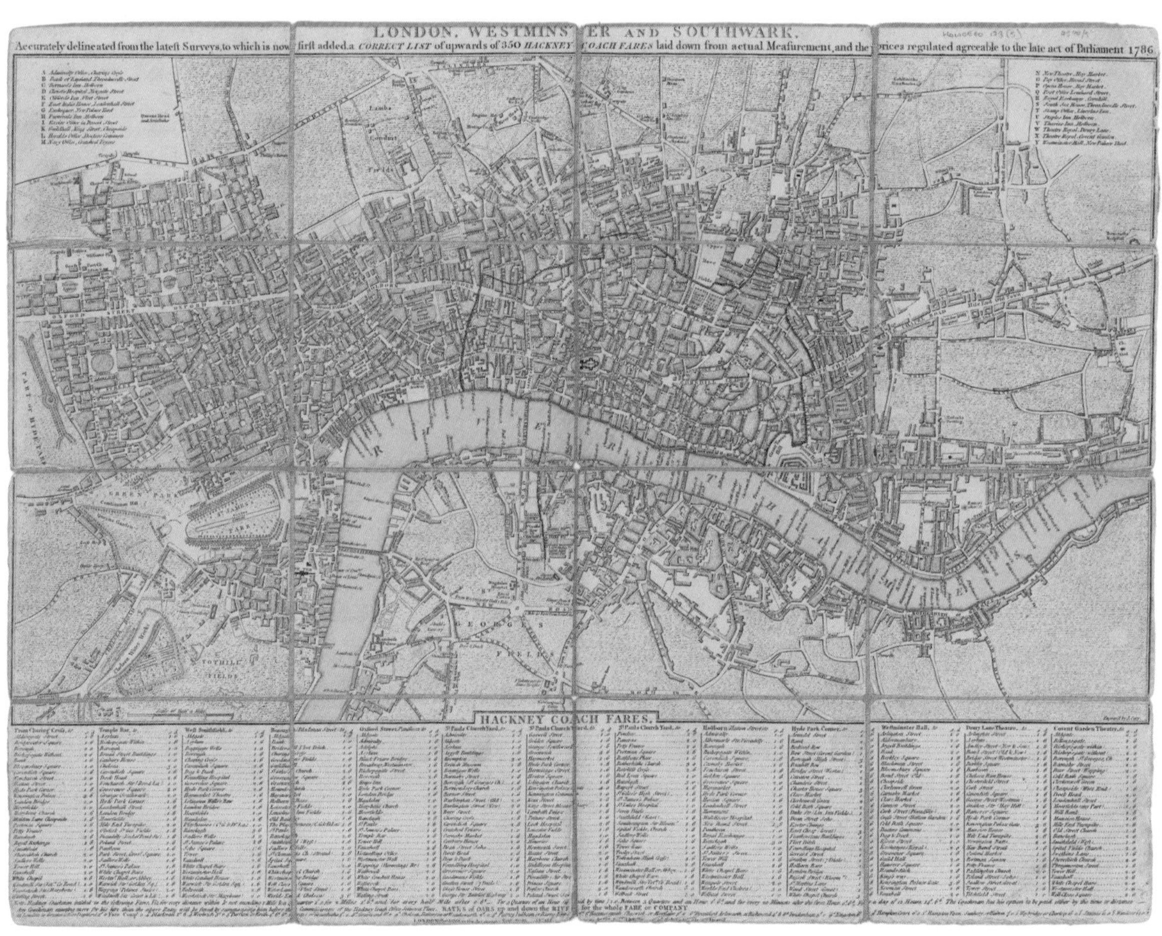

Father Thames

58 **FAIRBURN, John**

London and Westminster.

Publication
London, John Fairburn, 1795.

Description
Engraved map, partly hand-coloured and dissected into 16 sections, mounted on linen, in a custom-made marbled slipcase.

Dimensions
640 by 450mm. (25.25 by 17.75 inches).

Scale
$3\frac{3}{4}$ inches to 1 statute mile.

References
Howgego 203 (1).

The second appearance of Fairburn's map, engraved by Mogg and with an alphabetical list of streets and references in the bottom margin, with engraving of Father Thames top right.

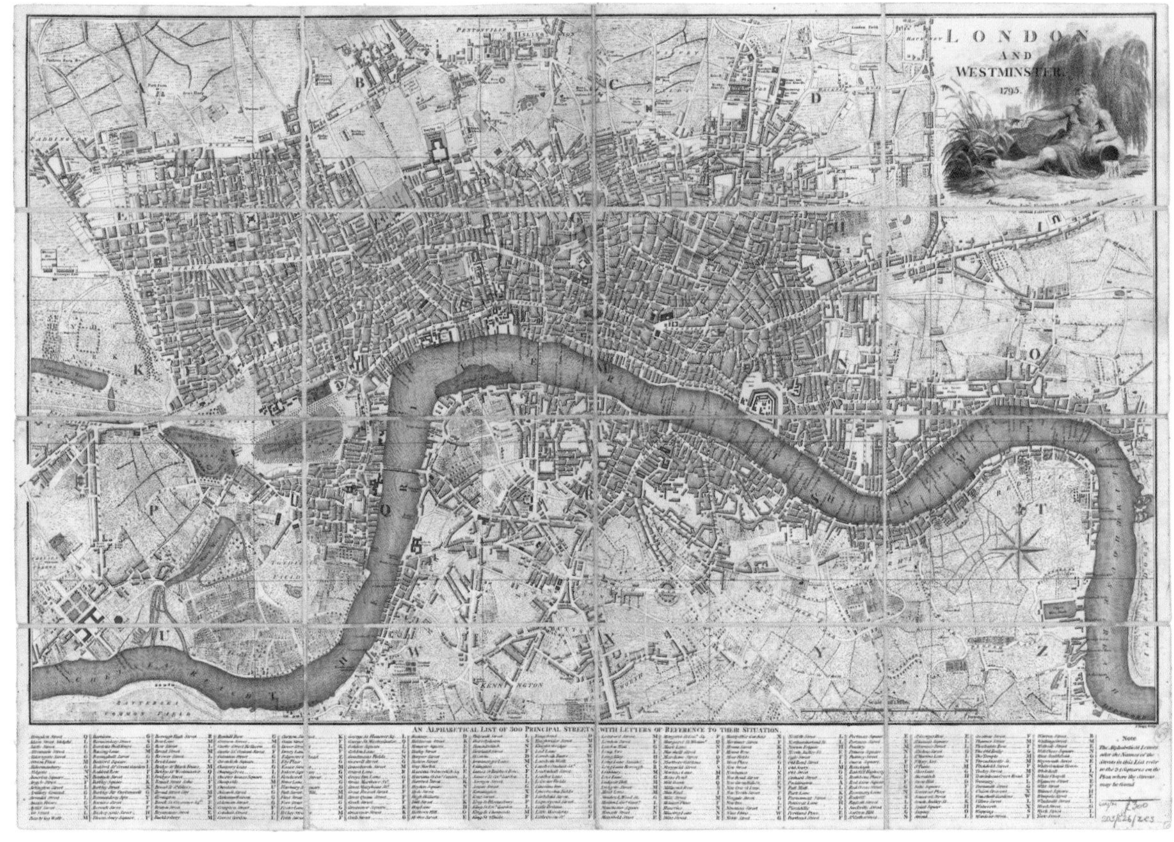

Plan showing military dispositions

59 **BOWLES, Carington**

Bowles's Reduced New Pocket Plan of the Cities of London and Westminster, with the Borough of Southwark, exhibiting the New Buildings to 1798.

Publication
London, Printed for Bowles and Carver at their Map & Print Warehouse. No. 69 St. Paul's Church Yard, 1798.

Description
Engraved plan, manuscript annotations with key below plan, a few old folds re-inforced with minor loss.

Dimensions
475 by 615mm (18.75 by 24.25 inches).

Scale
4 inches to 1 statute mile.

References
Howgego 158a (10a).

Below the plan is a nine column reference table listing churches and principal buildings, and an explanation of the plan's grid reference. The most important feature of this map is the additional colouring keyed in a manuscript note in the bottom margin. This shows the disposition of infantry (red) and cavalry (blue) in various parts of London. Infantry and cavalry stations and quarters are marked with a square, whereas posts and guards with a circle.

▪ a Cavalary Station and Quarter ▪ an Infantry Station and Quarter

• a Cavalary Post and Guard • an Infantry Post and Guard

The largest map ever printed in Georgian Britain

60 **HORWOOD, Richard**

Plan of the Cities of London and Westminster The Borough of Southwark and Parts Adjoining, Shewing Every House.

Publication
London, 1799.

Description
Folio (580 by 430mm), calligraphic title in an oval cartouche, engraved plan on eight large joined folding sheets with original hand colouring, dedication to the Phoenix Fire Office on sheet G4, occasional very light browning and offsetting as usual, a few minor repaired tears, all on wove paper, bookplate of A. G. E. Carthew, half calf, over marbled boards.

Dimensions
(if joined approx.) 4000 by 2280mm (157.5 by 89.75 inches).

Scale
26 inches to 1 statute mile.

References
Howgego 200 (1).

Horwood's map was produced for use by the Phoenix Fire Office and is dedicated to the Trustees and Directors. It was the largest map ever printed in Britain at the time, and the first attempt to produce a map of London with all of the houses delineated and numbered, an invaluable aid for the insurance office and very useful in identifying the street numbering of eighteenth century London. The numbering of buildings did not begin until about 1735, when the practice of identifying a building by describing it as 'by', 'opposite' or 'over against' some other building was recognized as confusing and erroneous. However, even by the date of printing of Horwood's map, it was still not universal.

The map is produced to exactly the same scale as the Rocque map of fifty years before and so enables us to compare the development of London on a street-by-street basis in the second half of the eighteenth century. It is also fascinating to look at the fringes of London and see the areas that were soon to be swallowed up. Thomas Lord's cricket ground is shown in its original place, in what is now Dorset Square. It was opened there in 1787 (7 years before the relevant sheet on the map was published). The pressure of London's development led to a rent increase in 1809 that resulted in Lord moving his cricket ground to the greener pastures of St. John's Wood.

Very little is known about Horwood (1758–1803). Most likely he was working for the Phoenix Assurance Company on surveying jobs when he began the enormous task of surveying the whole of the built-up area of London. "According to his own account the prepration of the plan gave him nine years' severe labour and he himself "took every angle, measured almost every line, and after that plotted and compared the whole work"". He sent a small sample of the plan showing Leicester Square and its neighborhood to all the London vestries with a letter promising that those "who gave him Encouragement" could have a "compleat" copy by "the year 1792". His estimate proved to be over-optimistic and only one sheet – B2 (Grosvenor Square-Piccadilly) – was published by 1792" (Howgego, p.22). In January 1798 he wrote to the Phoenix Assurance Company offering to dedicate his map to the company if the directors would make him a loan of £500 to enable him to finish the work. His request was granted but this, in addition to an award from the Society of Arts, were too little and too late and, in 1803, Richard Horwood died in Liverpool in poverty and obscurity, so sharing the fate of other great men like John Stow and Wenceslaus Hollar, to whom London had failed to honour her debt of gratitude.

THAMES
THE RIVER
SOUTHWARK
BERMONDSEY
BERMON
WALWORTH
THE TOWER

St. MARY LE BONE
PARISH
SAINT GEORGE'S
PARISH of St. MARTINS
PARISH
THE RIVER
THAMES

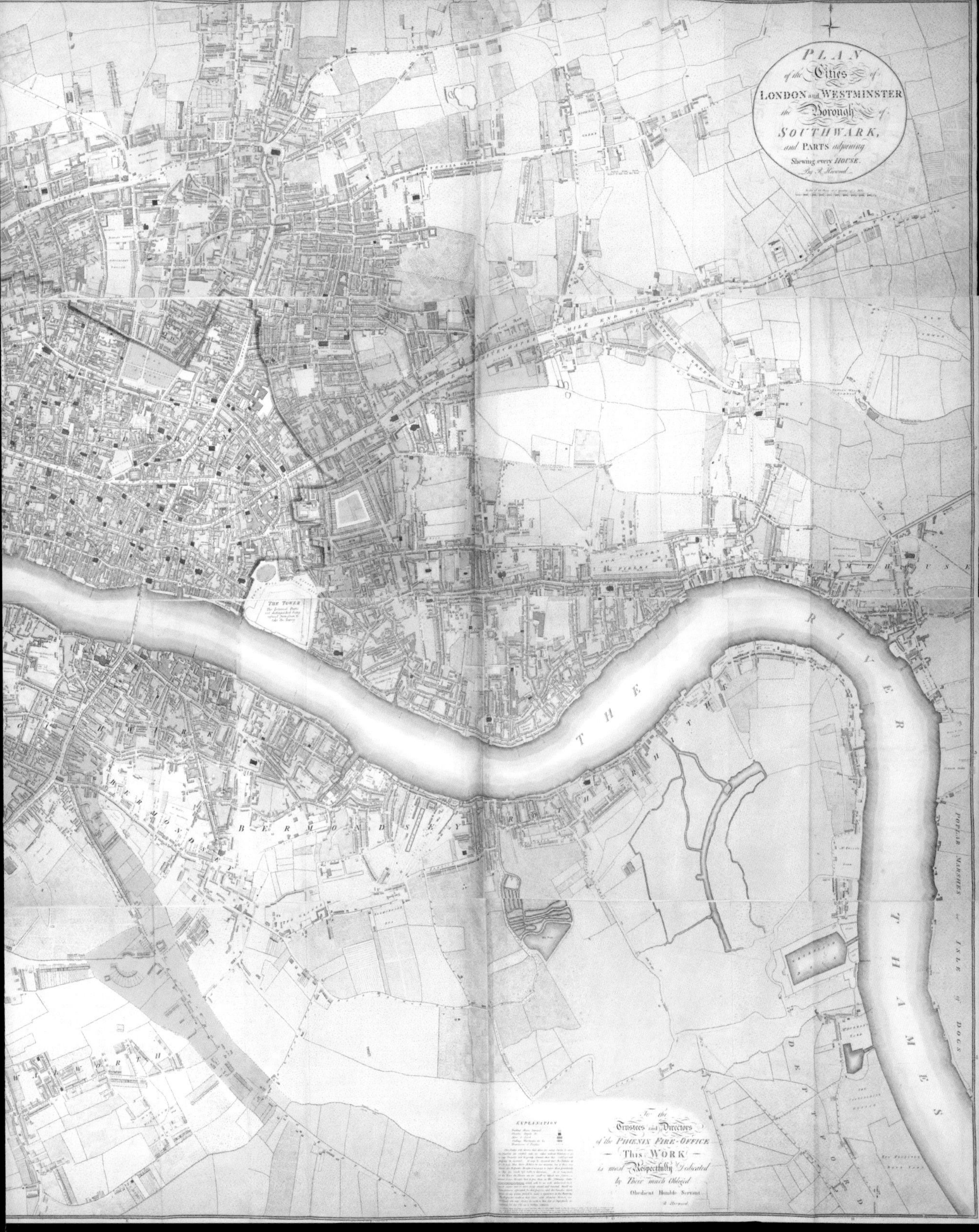
PLAN
of the Cities of
LONDON and WESTMINSTER
the Borough of
SOUTHWARK,
and PARTS adjoining
Shewing every HOUSE.
By R. Horwood.
MILE END OLD TOWN
THE TOWER
SOUTHWARK
BERMONDSEY
ROTHERHITHE
LIMEHOUSE
THE RIVER THAMES
POPLAR MARSHES or ISLE of DOGS
DEPTFORD
EXPLANATION
To the
Trustees and Directors
of the PHŒNIX FIRE-OFFICE
This WORK
is most Respectfully Dedicated
by Their much Obliged
Obedient Humble Servant
R. Horwood

Bedford Square and London Docks

61 **FADEN, William**

A Plan of the Cities of London & Westminster and the Borough of Southwark...

Publication
London, Published by Wm. Faden, Charing Cross, Feby. 7th, 1785. [but c.1800].

Description
Engraved plan, tear to foot of centre fold skilfully repaired.

Dimensions
590 by 800mm (23.25 by 31.5 inches).

Scale
5¼ inches to 1 statute mile.

References
Howgego 180.

To the upper right of the plan three putti raise a curtain to reveal the title, below reclines a personification of the Thames. To the plan itself cross-hatching and stippling are used to distinguish land use, with differentiation made between fields and parkland. The plan would seem to be slightly later than the imprint date, as Bedford House has been demolished and replaced by Bedford Square, which did not occur until 1799, and a large part of the London Docks have been constructed, work upon which did not commence until 1799.

Separately issued plan of the proposed West India Docks

62 **WALKER, Ralph**

Plan of the West India Wet Docks...

Publication
[London], Published as the Act directs, by R. Walker, Blackwall, Jany. 1, 1801.

Description
Hand-coloured engraved plan, key and title below plan, engraved view of the north side of the docks above plan.

Dimensions
570 by 840mm (22.5 by 33 inches).

Scale
Approx 32 inches to 1 statute mile.

References
BLMC Maps 3500.(9.).

The plan shows the intended docks together with the rope grounds, paths, streams, and fields that would be obliterated by the new works.

The plan is dedicated to George Hibbert and Robert Milligan, two powerful West Indies merchants who lobbied Parliament in order to construct a secure dock to load and off load their cargoes. The construction of the docks was authorised by the West India Docks Act of 1799. They would be completed by 1802, with the canal to the south being finished in 1805.

Ralph Walker (1749–1824) was appointed Resident Engineer to the West India Dock Company in August 1799 on a salary of £600 p.a. Together with William Jessop, the chief engineer, he was responsible for the construction of the docks. However, he would resign from his post in October of 1802, following a professional disagreement with Jessop over a structural failure at the docks.

Rare. OCLC records only a single example: that of the British Library.

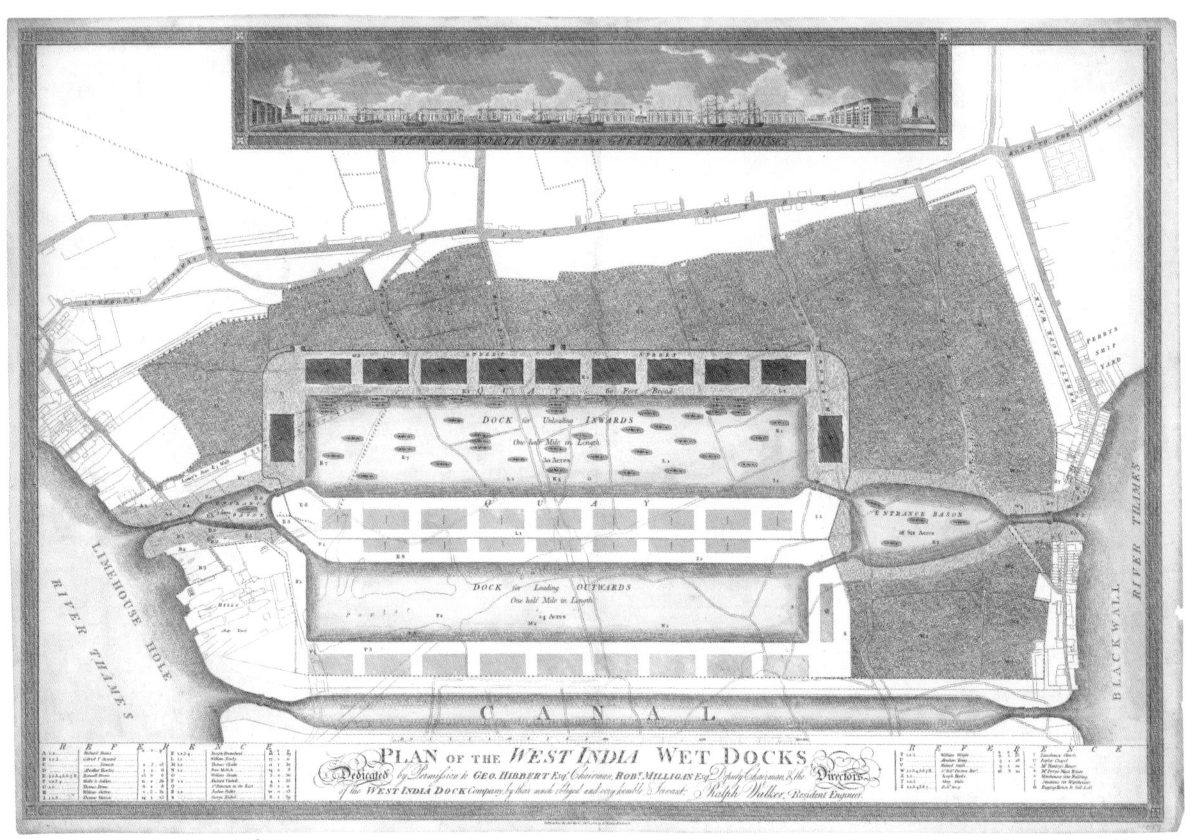

Plan of London highlighting the newly opened Somerset House

63 **LAURIE, Robert and James WHITTLE**

A New Plan of London, Westminster and Southwark.

Publication
London, Engraved by S.W. Cooke. Published 12th Octr. by Laurie & Whittle, 53, Fleet Street, London, 1801.

Description
Engraved map, original hand-colour, some spotting to margins and key.

Dimensions
500 by 675mm (19.75 by 26.5 inches).

Scale
3¾ inches to one statute mile.

References
Howgego 223 (1).

Below the map is a list of Public Offices and Buildings. The list draws specific attention to Public Offices situated in Somerset Place (Somerset House). By 1801, Somerset House was nearing completion, with many of the Public Offices having moved in. Its construction came about due to growing criticism that London had no great public buildings to compare with those of continental Europe, most notably Paris. To address this, an Act of Parliament was passed in 1775 to establish "Publick Offices in Somerset House, and for embanking Parts of the River Thames lying within the bounds of the Manor of Savoy". In the same year Sir William Chambers was commissioned to design and construct the building. The building would consume the rest of his life, being completed after his death by James Wyatt. The plan lists the numerous offices that the building housed:

"The Navy Pay, Victualling, Sick and Hurt, Stamp, Tax, Dutchy of Lancaster & Cornwall, Auditors of Public Accounts, Pipe, Lord Treasurer's, Remembrancers, Hackney Coach, Chair, Cart, Hawkers, Privy-Seal, Salt, Stage Coach, & Horse Duty Offices. Royal Society of Arts, Royal and Antiquarian Society".

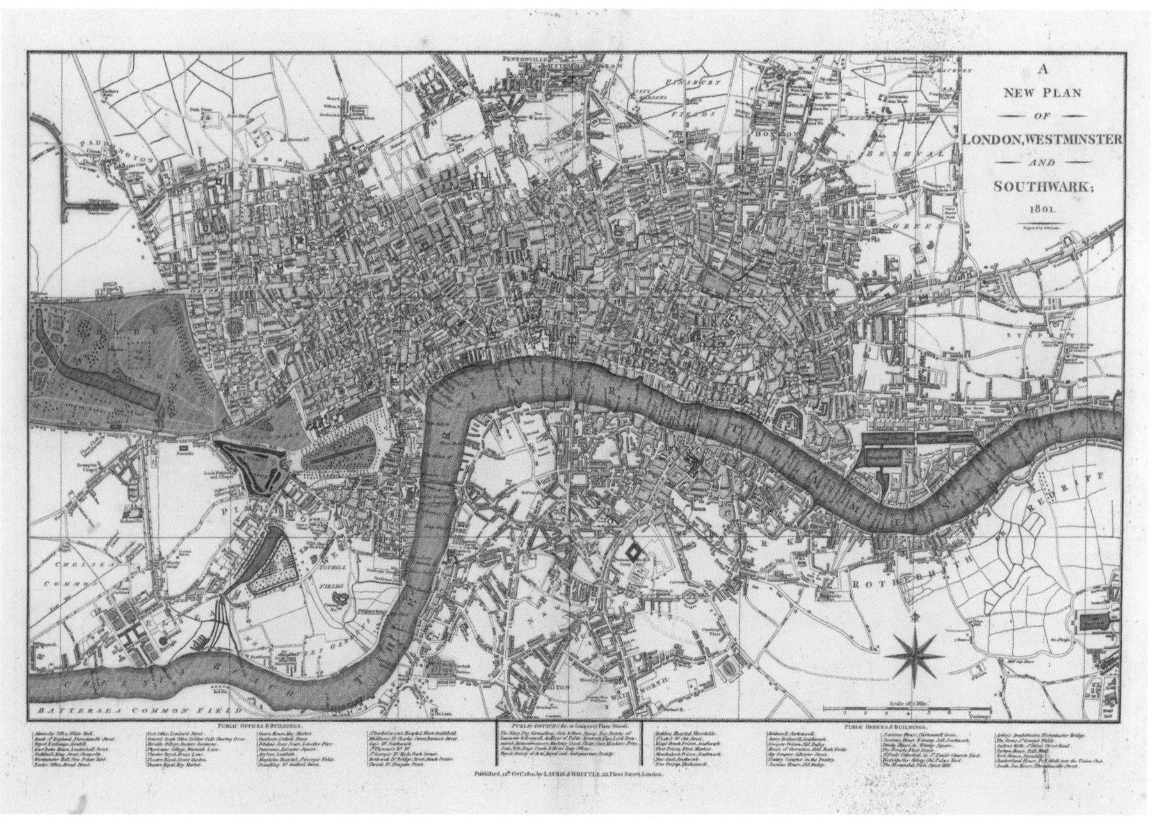

Plan of London showing troop dispositions

64 **FADEN, William**

A New Pocket Plan of the Cities of London and Westminster: with the borough of Southwark: Comprehending the New Buildings and Alterations to the Year 1803.

Publication
London, Published by W. Faden, Geographer to the King and to the Prince of Wales, Charing Cross, 1803.

Description
Engraved plan on two sheets, original outline hand-colour, manuscript annotations to map with manuscript key to right margin, a few old tears skilfully repaired and old folds reinforced.

Dimensions
465 by 1030mm (18.25 by 40.5 inches).

Scale
6¼ inches to 1 statute mile.

References
Howgego 186 (8).

The plan gives information on parishes in Westminster, Surrey, and Middlesex, with the Bills of Mortality. A coloured key shows the limits of the City of London, the intended buildings or new streets not finished, the Liberties of Westminster, and Rules of the Bench and Fleet. The border is divided into miles and furlongs based on St Paul's, and intended and unfinished development includes the area east of the Edgware Road and south of Euston Road; a circus at the east end of the Strand; and the London Docks. However, the most important feature of this map is the additional coloured key in a manuscript note to the right hand margin. This shows the disposition of troops, regular and volunteer, in various parts of London. Pencilled numbers appear against many of the sites but what these denote is unclear.

"This Ground belongs to the London Dock Company the Plan not known"

65 **MOGG, Edward**

An Entire New Plan of the Cities of London & Westminster; with the Borough of Southwark: Comprehending the New Buildings and other Alterations to the Year 1805.

Publication
London, Publd. by Edwd. Mogg, No.14 Little Newport Street, Leicester Square, 1805.

Description
Engraved plan, original outline hand-colour, dissected and mounted on linen.

Dimensions
890 by 460mm (35 by 18 inches).

Scale
6¼ inches to 1 statute mile.

References
Howgego 227 (3).

The plan shows the limits of the City of London in red; the Rules of the King's Bench, and Fleet Prisons – in which some prisoners were allowed to reside – in yellow; parks and squares highlighted in green; and proposed streets and buildings marked in outline – such as Bloomsbury and Marylebone. The second phase of the London Docks is marked with the words: "This Ground belongs to the London Dock Company the Plan not known".

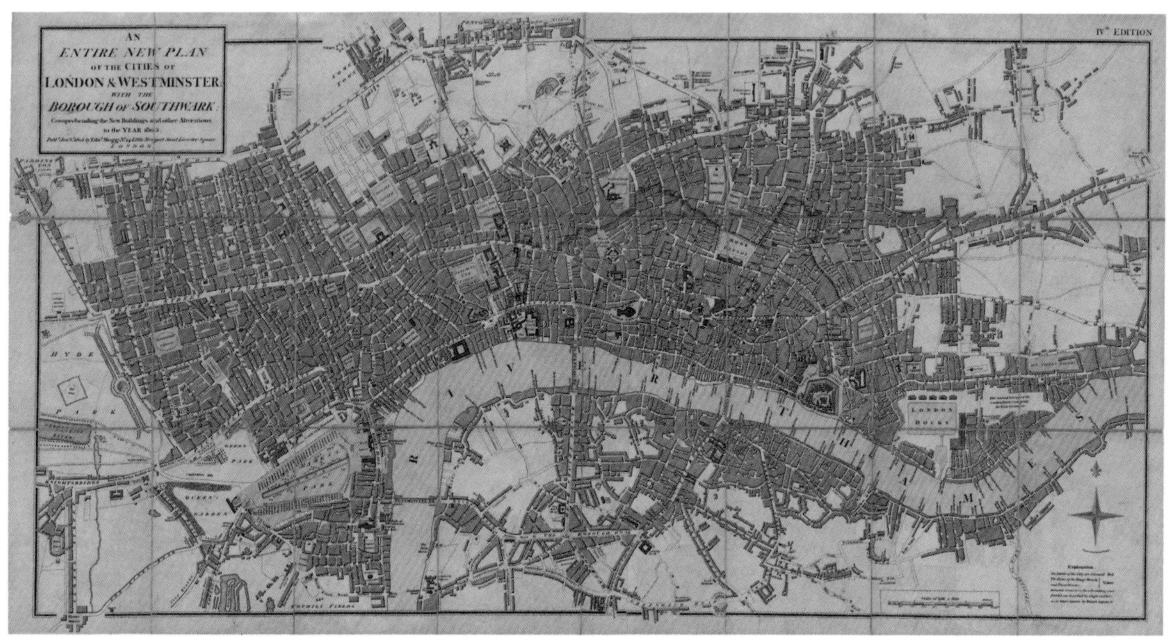

115 churches, 128 public buildings, and 36 squares…

66 **STRATFORD, James**

London, extending from the Head of the Paddington Canal West, to the West India Docks East: with the proposed Improvements between the Royal Exchange & Finsbury Square.

Publication
London, Publish'd as the Act directs by J. Stratford, No. 112, Holborn Hill, Octr. 6th, 1806.

Description
Engraved plan, reference table below, right margin with loss, a few nicks and tears to margins, a few affecting image, skilfully repaired.

Dimensions
340 by 575mm (13.5 by 22.75 inches).

Scale
3⅜ inches to one statute mile.

References
Howgego 240.

The plan is unusual in that it gives its exact extent in the title. Below the plan is a table listing 115 churches, 128 public buildings, and 36 squares. It was published in 'London, being an accurate history & description of the British Metropolis… by David Hughson'.

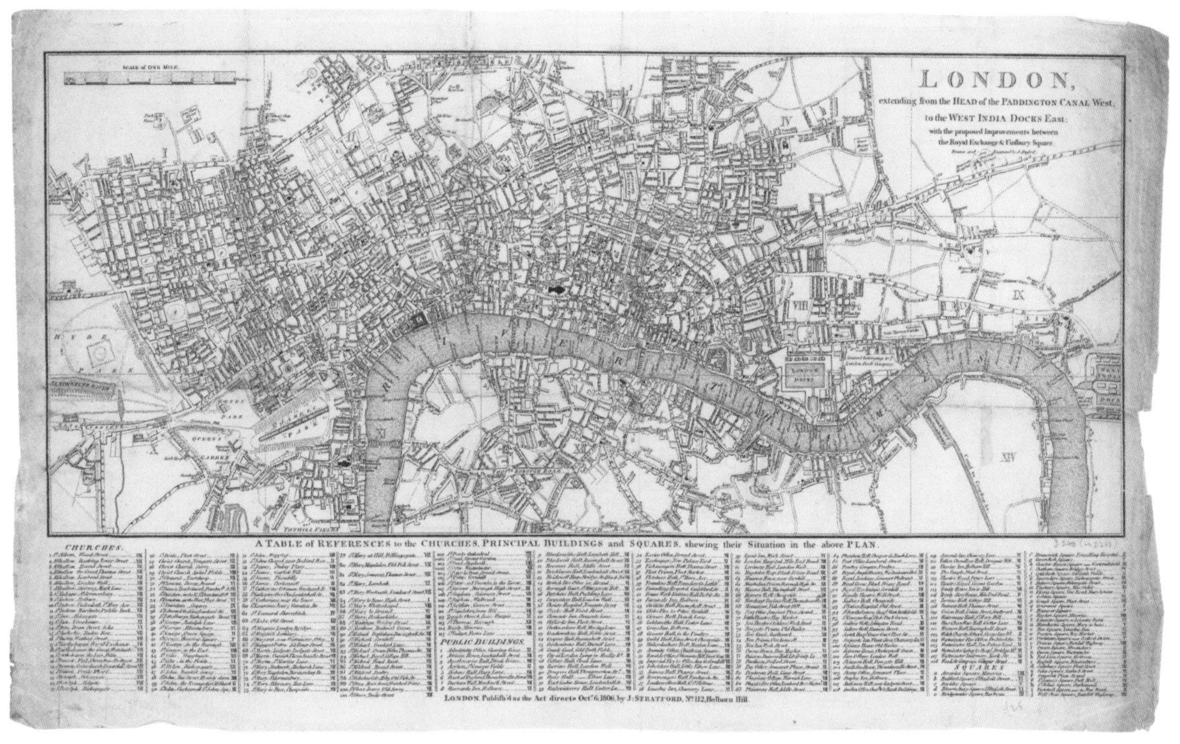

A picture of London

67 **PHILLIPS, Richard**

A Plan of London, with its Modern Improvements.

Publication
[London], Published by Ricd. Phillips New Bridge St, May 1st 1808

Description
Engraved plan, minor wear to folds.

Dimensions
265 by 642mm (10.5 by 25.25 inches).

Scale
Approx. 3½ inches to 1 statute mile.

References
Howgego 229 (4).

Published in a 'Picture of London for 1809... Printed for Richard Phillips'. According to Howgego this state of the map also appeared in the 1810, 1811, 1812 and 1813 editions of the 'Picture of London'.

London docks and the Croydon Canal

68 **STOCKDALE, John**

A New Plan of London, XXIX Miles in Circumference.

Publication
London, Printed by J. Stockdale, Piccadilly, Jany. 2nd 1797 [but c.1809].

Description
Engraved map on four sheets, a few tears to old folds skilfully repaired.

Dimensions
1050 by 1500mm (41.25 by 59 inches).

Scale
6½ inches to one statute mile.

References
Howgego 213 (2).

This large and detailed plan of London is very similar stylistically to much of the output of John Rocque, although Stockdale is somewhat sparser, with few baroque flourishes. The plan is dated 1797, yet it was clearly issued some twelve years later as it not only shows the West India Docks (opened in 1802), the London Docks (opened in 1805), and the East India Docks (opened in 1806), but also the Croydon Canal, which was not in use until 1809. Also shown are the intended routes, marked by dotted lines, of Regent's Bridge (later renamed Vauxhall Bridge), and Waterloo Bridge.

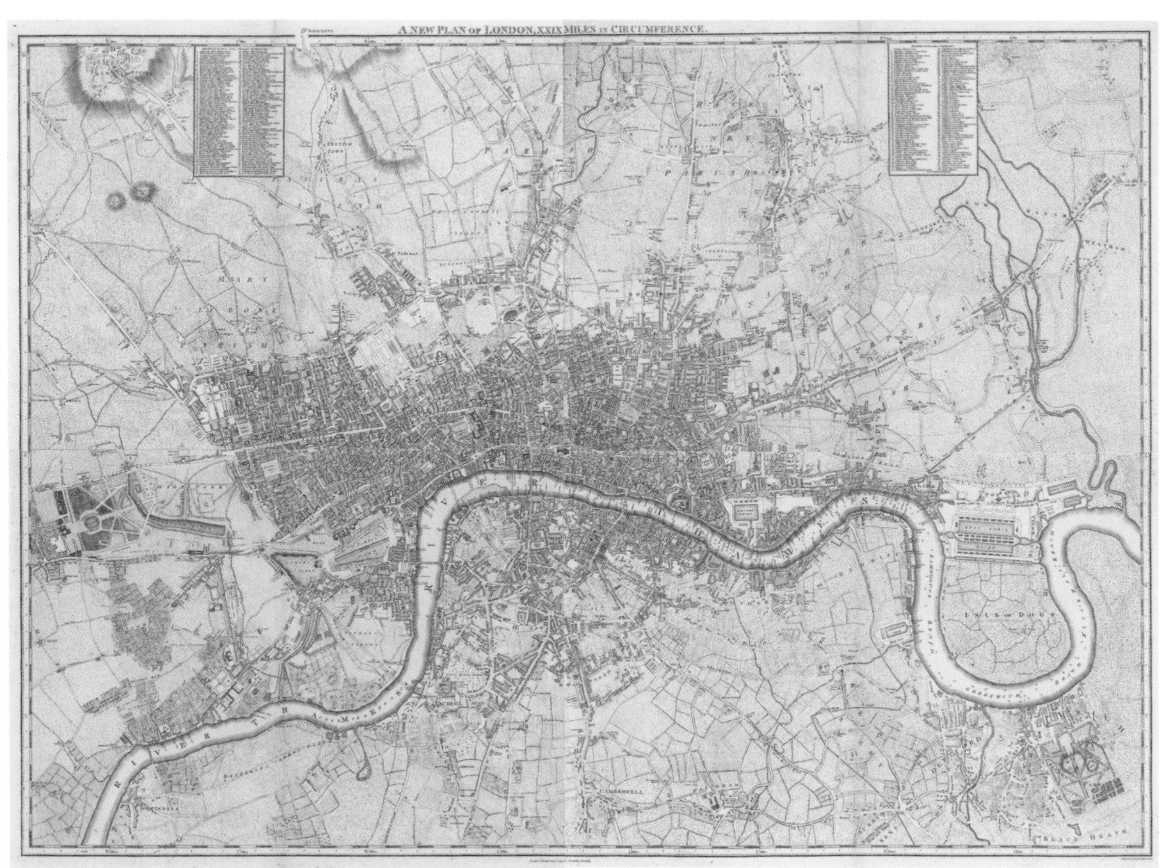

A rare and detailed plan of late Georgian London

69 **LANGLEY, Edward, & William BELCH**

Langley and Belch's New Map of London.

Publication
London, Published as the Act directs, by Langley & Belch, No.173 High St. Borough, May 1st, 1812.

Description
Engraved plan, fine original hand-colour, dissected and mounted on new linen, 24 vignette views of London landmarks above and below plan, some even age-toning, folding into original slip case, rubbed and scuffed.

Dimensions
545 by 790mm (21.5 by 31 inches).

Scale
3¾ inches to 1 statute mile.

References
Howgego 256 (1).

The plan shows the intended construction of two bridges across the Thames: that at Vauxhall, opened in 1816, and the other near the Strand, here named Strand Bridge but upon its opening in 1817 renamed Waterloo Bridge. The border is marked off in half miles, with the plan itself divided into lettered squares for use with Langley and Belch's 'Companion to their new map of London'.

Above and below the plan are 24 views of London landmarks; they are in order from upper left to lower right: St. Paul's Cathedral, Westminster Bridge, Bank, Lambeth Palace, Royal Exchange, Greenwich Hospital, Monument, Chelsea Hospital, Charing Cross, the Tower of London, Horse Guards, London Bridge, Westminster Abbey, Westminster Hall, The Admiralty, India House, Mansion House, the London Docks, West India Docks, East India Docks, The Queen's Palace in St. James's Park, Blackfriars Bridge, Somerset House, and the Guildhall.

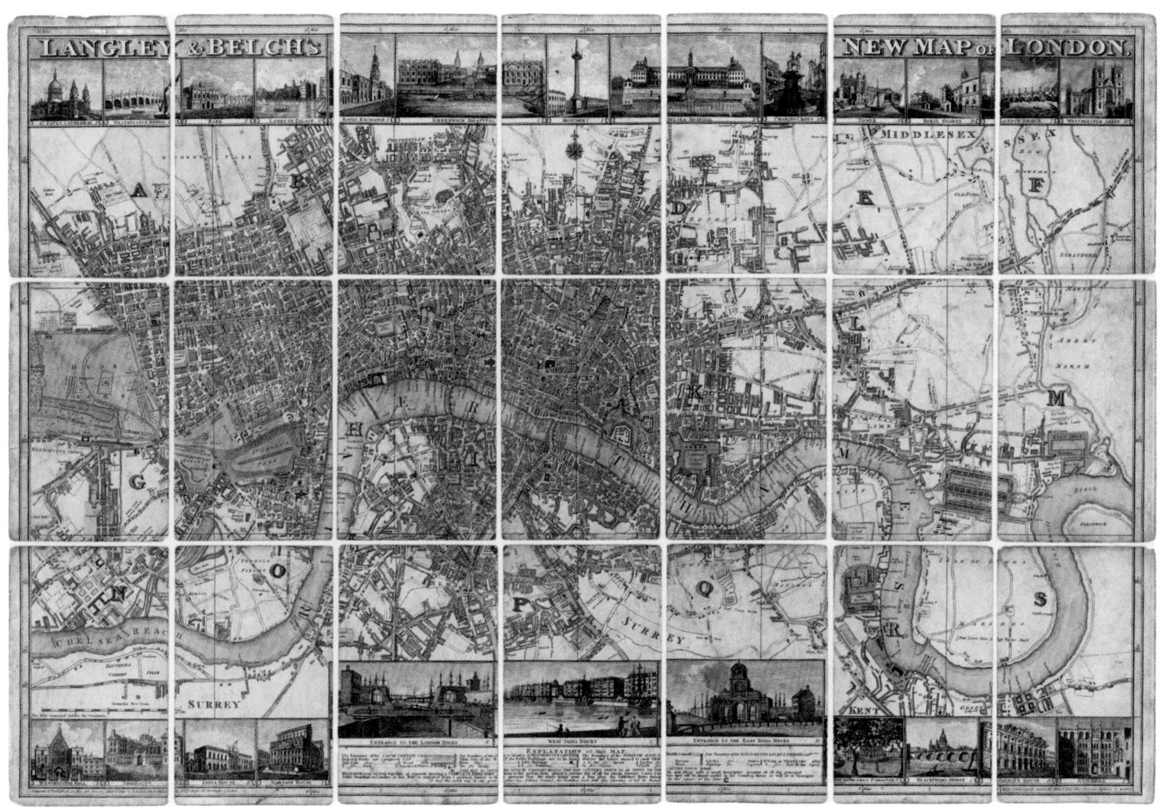

A reverse merger…

70 **BELCH, William**

New Map of London.

Publication
London, Printed & Published by W. Belch, No. 1 Staverton Row. Newington Butts, July 1st, 1820.

Description
Hand-coloured engraved plan, dissected and mounted on linen, housed within modern marbled cardboard slip-case.

Dimensions
545 by 795mm (21.5 by 31.25 inches).

Scale
3¾ inches to 1 statute mile.

References
Howgego 256 (4).

Further description in lower central panel reads: "This map is divided into 18 compartments, the red letters serve as an index to the letter press book, which book may be had together or separate forming a Complete Directory down to the present time giving a correct list of all the Streets Squares Lanes Courts and Alleys with a small descriptive account of all the principal Public Buildings in London as this Map is engraved from a personal Survey to the present time, the whole being upon a New plan the Publishers flatter themselves they will be found usefull in the Counting House as well as to Strangers."

This attractive map includes twenty-four views of London buildings within the top and bottom borders. The map shows the land east of London Docks cleared ready for further expansion and the lands north and west of Tavistock Square ready for development.

Originally published in 1812 under the imprint of Edward Langley and William Belch, who appear to have traded under separate names from 1820 onwards, but still gave a joint dividend in 1824.

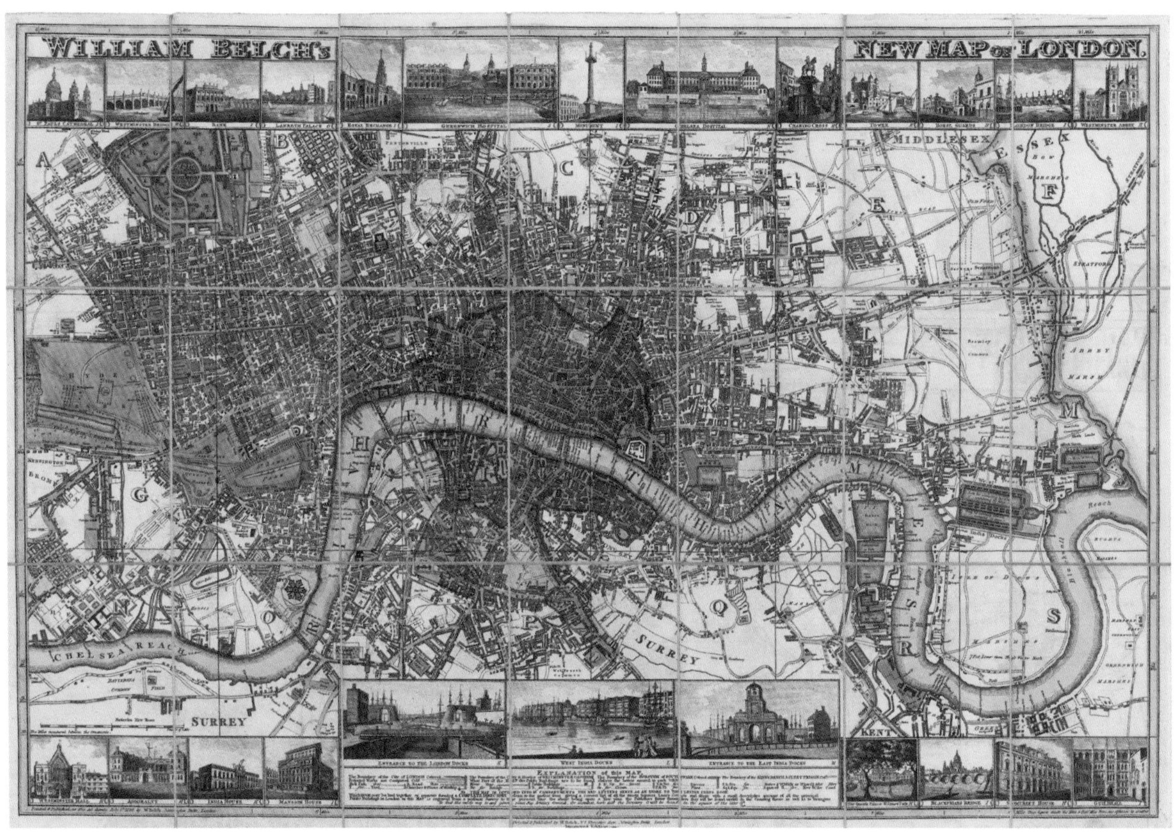

The proposed Regent Street

71 **BASIRE, James**

[Regent Street] Plan,presented to the House of Commons, of a Street, proposed from Charing Cross to Portland Place, leading to the Crown Estate in Mary-le-Bone Park, on a reduced scale. Ordered by the House of Commons.

Publication
London, Luke Hansard & Sons, 10th May, 1813.

Description
Large engraved plan of the proposed Regent Street and the surrounding area with the streets highlighted in colour, watermark 1812.

Dimensions
534 by 330mm (21 by 13 inches).

Scale
16½ inches to 1 statute mile.

Large engraved plan of the proposed Regent Street and the surrounding area with the streets highlighted in colour.

The title appears along the top, with a descriptive note below the plan and a scale bar at the bottom centre. Crown Property is highlighted in blue.

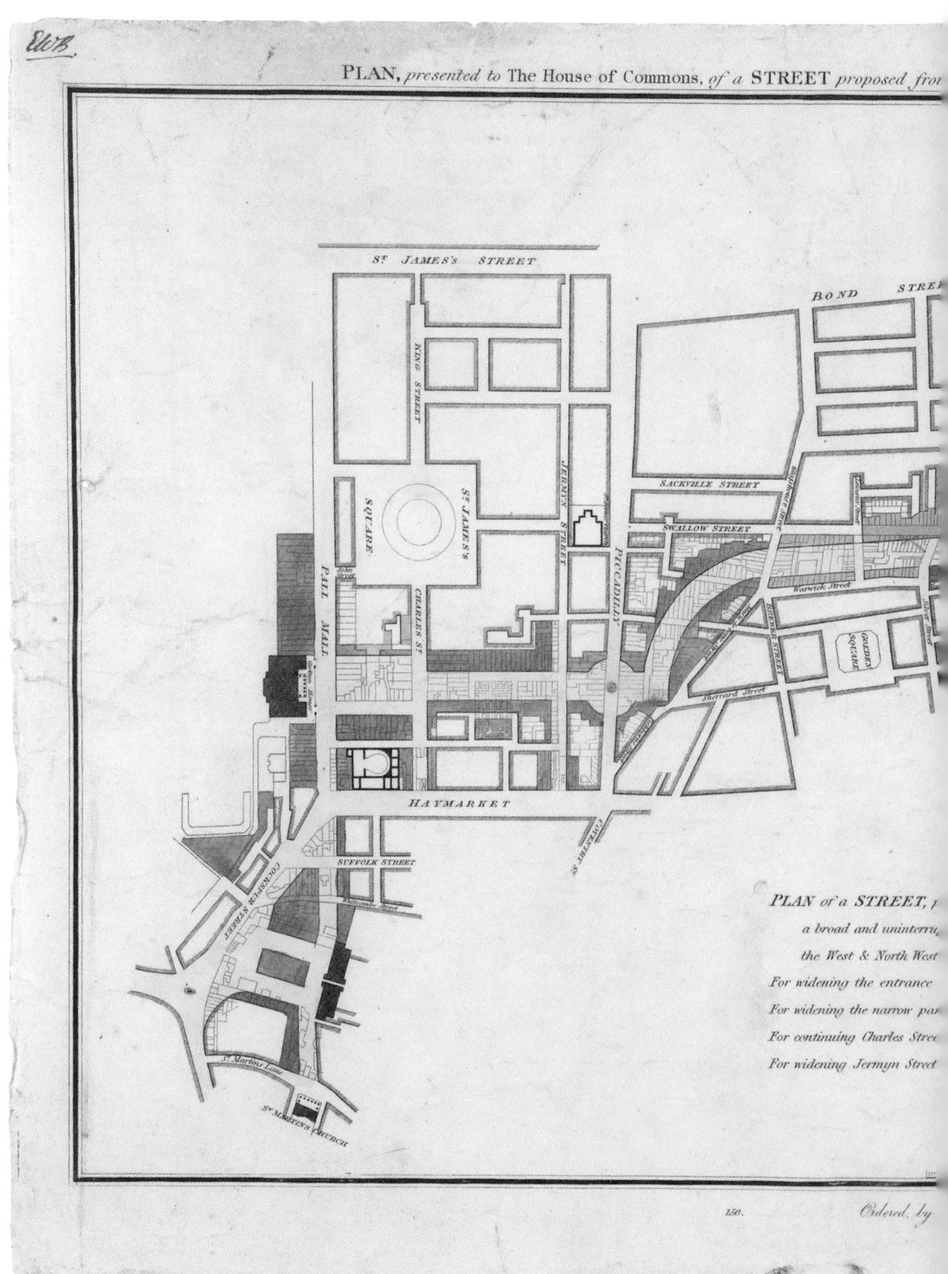

Starting at Carlton House, Regent Street ran through crownland at Piccadilly (where a circus was built) before turning north-west along Swallow Street, in Soho, finally joining Portland Place north of Oxford Street. The title appears along the top, with a descriptive note below the plan and a scale bar at the bottom centre. Regent Street was designed by John Nash and built between 1813 and 1823.

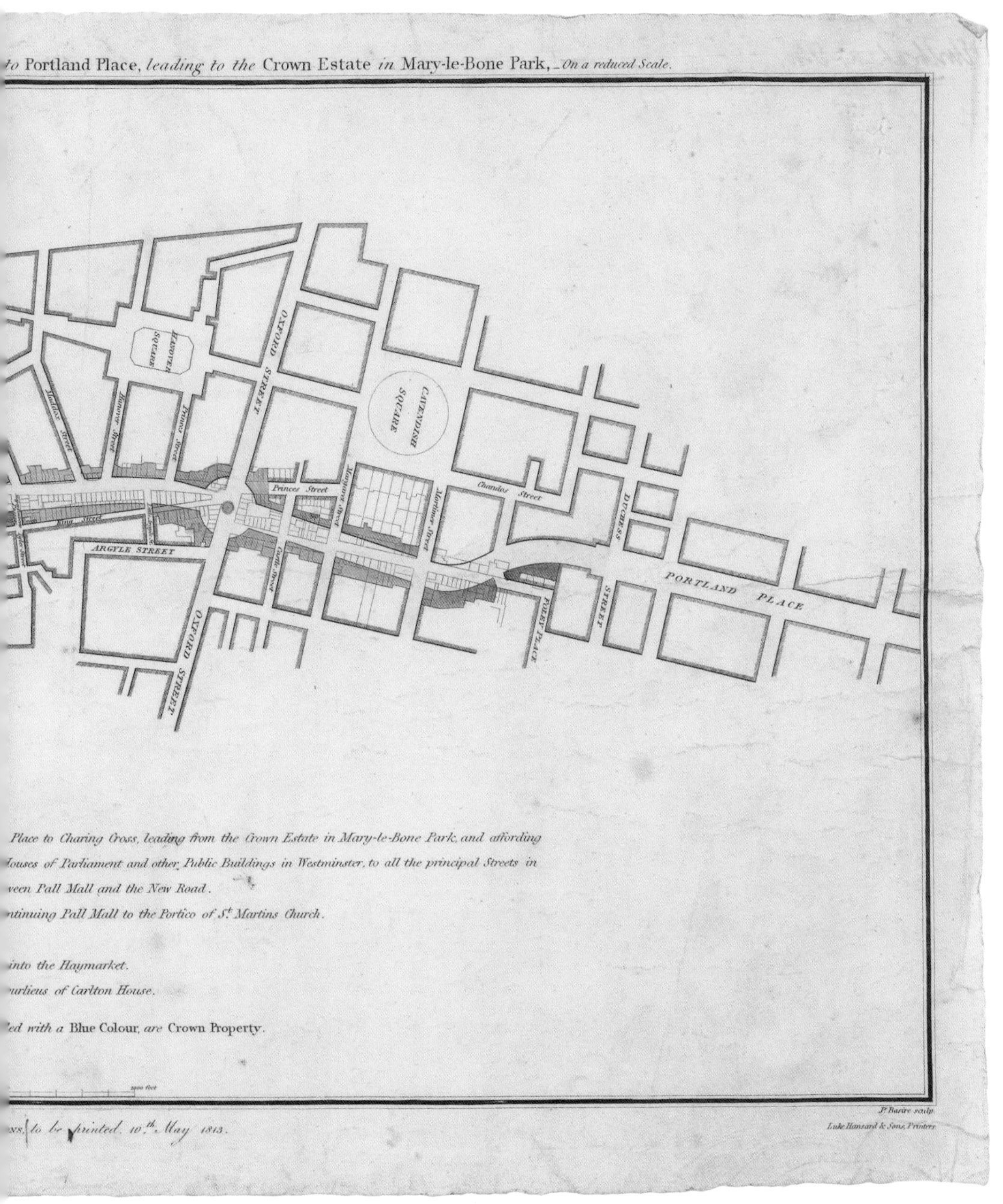

Greenwood's large-scale map of Middlesex

72 **GREENWOOD, Christopher**

Map of the County of Middlesex from an Actual Survey made in The Years 1818 & 1819. By C. Greenwood. To the Nobility, Clergy & Gentry of Middlesex. This Map of the County Is most respectfully Dedicated by The Proprietors.

Publication
London, Published by the Proprietors G. Pringle and C. Greenwood, No. 50, Leicester Square, Oct. 25, 1819.

Description
Engraved map on four sheets, view of London from Buckingham House, upper left, stab bound.

Dimensions
1400 by 1120mm. (55 by 44 inches).

Scale
2 inches to 1 statue mile.

The maps by Christopher and John Greenwood set new standards for large-scale surveys. Although they were unsuccessful in their stated aim to map all the counties of England and Wales, it is probably no coincidence that of the ones they missed, Buckinghamshire, Cambridgeshire, Herefordshire, Hertfordshire, Norfolk and Oxfordshire, all except Cambridgeshire were mapped by Andrew Bryant in a similar style and at the same period. From a technical point of view, the Greenwoods' productions exceeded the high standards set in the previous century, though without the decoration and charming title-pieces that typified large-scale maps of that period.

The Greenwoods started in 1817 with Lancashire and Yorkshire and by 1831 they had covered 34 counties. Their maps were masterpieces of surveying and engraving techniques, and in view of the speed at which they were completed, their accuracy is remarkable. They mark the boundaries of the counties, hundreds and parishes, churches and chapels, castles and quarries, farmhouses and gentlemen's seats, heaths and common land, woods, parliamentary representatives and distances between towns. The price of 3 guineas each compares with the first edition Ordnance Survey sheets of 7s 6d, though the latter did not relate to complete counties.

TO THE NOBILITY, CLERGY & GENTRY OF MIDDLESEX.
This Map of the County
Is most respectfully Dedicated by
THE PROPRIETORS.
H E R T F O R D S H I R E
B U C K I N G H A M S H I R E
E S S E X
K E N T
S U R R E Y
G O R E H U N D R E D
E L T H O R N E H U N D R E D
E D M O N T O N H U N D R E D
I S L E W O R T H H U N D R E D
S P E L T H O R N E H U N D R E D
RIVER THAMES
Map
OF
the County of
MIDDLESEX.
from an Actual Survey made in the Years 1818 & 1819
By
C. GREENWOOD.
London
Published by the Proprietors
G. PRINGLE and C. GREENWOOD,
EXPLANATION.

Panorama keysheet

73 *Panorama de Londres.*

Publication
[?Paris, c. 1820].

Description
Engraved circular panorama with individual buildings numbered, and key in French and English, horizontal and vertical folds.

Dimensions
450 by 345mm (17.75 by 13.5 inches).

The keysheet was produced to accompany an explanatory leaflet for Robert and Henry Aston Barker's 'Panorama of London painted as if viewed from the roof of Albion Mills on the South Bank', and covers St. Paul's, the Tower, Westminster Bridge, Whitehall, Leverian Museum, and the Temple in a circular view. Slight misspellings of the place names indicate that this plate, which was to accompany a brochure explaining a large cylindrical painted panorama of the capital, was produced in France.

Panoramic paintings became popular in the nineteenth century. The word "panorama", from the Greek "pan" (all) and "horama" (view), was coined by the Irish painter Robert Barker in 1792 to describe his view of Edinburgh painted on a cylindrical surface. Barker's inability to bring the image to a full 360 degrees disappointed him. To realize his true vision, Barker and his son, Henry Aston Barker, took on the task of painting a scene of London from the roof of the Albion Mills on the South Bank. Barker's accomplishment involved sophisticated manipulations of perspective not encountered in the panorama's predecessors, the wide-angle "prospect" of a city familiar since the sixteenth century, or Wenceslas Hollar's "long view" of London, etched on several contiguous sheets. When Barker first patented his technique in 1787, he had given it a French title: La Nature à Coup d' Oeil ("Nature at a glance"). In 1793 Barker moved his panoramas to the first purpose-built panorama building in the world, in Leicester Square, and made a fortune. Viewers flocked to pay a stiff three shillings to stand on a central platform under a skylight, and get an experience that was "panoramic" (an adjective that didn't appear in print until 1813). Visitors to Barker's Panorama of London could purchase a series of six prints that modestly recalled the experience; end-to-end the prints stretched 3.25 metres. In contrast, the actual panorama spanned 250 square metres. Due to the immense size of the panorama, patrons were given orientation plans, such as the present example, to help them navigate the scene.

Barker made many efforts to increase the realism of his scenes. To fully immerse the audience in the scene, all borders of the canvas were concealed. Props were also strategically positioned on the platform where the audience stood and two windows were laid into the roof to allow natural light to flood the canvases. Barker's Panorama was hugely successful and spawned a series of "immersive" panoramas: the Museum of London's curators found mention of 126 panoramas that were exhibited between 1793 and 1863.

1. S.t Paul's Church.
2. Bow Church.
3. Mansion House.
4. The Monument.
5. London Bridge.
6. The Tower.
7. The Harbour.
8. The Glass-Houses.
9. Christ Church.
10. Surry Street.
11. Leverian Museum.
12. Westminster Abbey.
13. Parliament House.
14. Westminster Bridge.
15. Horse Guards.
16. White-hall.
17. Sommerset House.
18. Albion Square.
19. River Thames.
20. Drury-lane Theatre.
21. The Temple.
22. Black Friar,s Bridge.
23. S.t Bridget's Church.
24. Surgeon's-Hall.

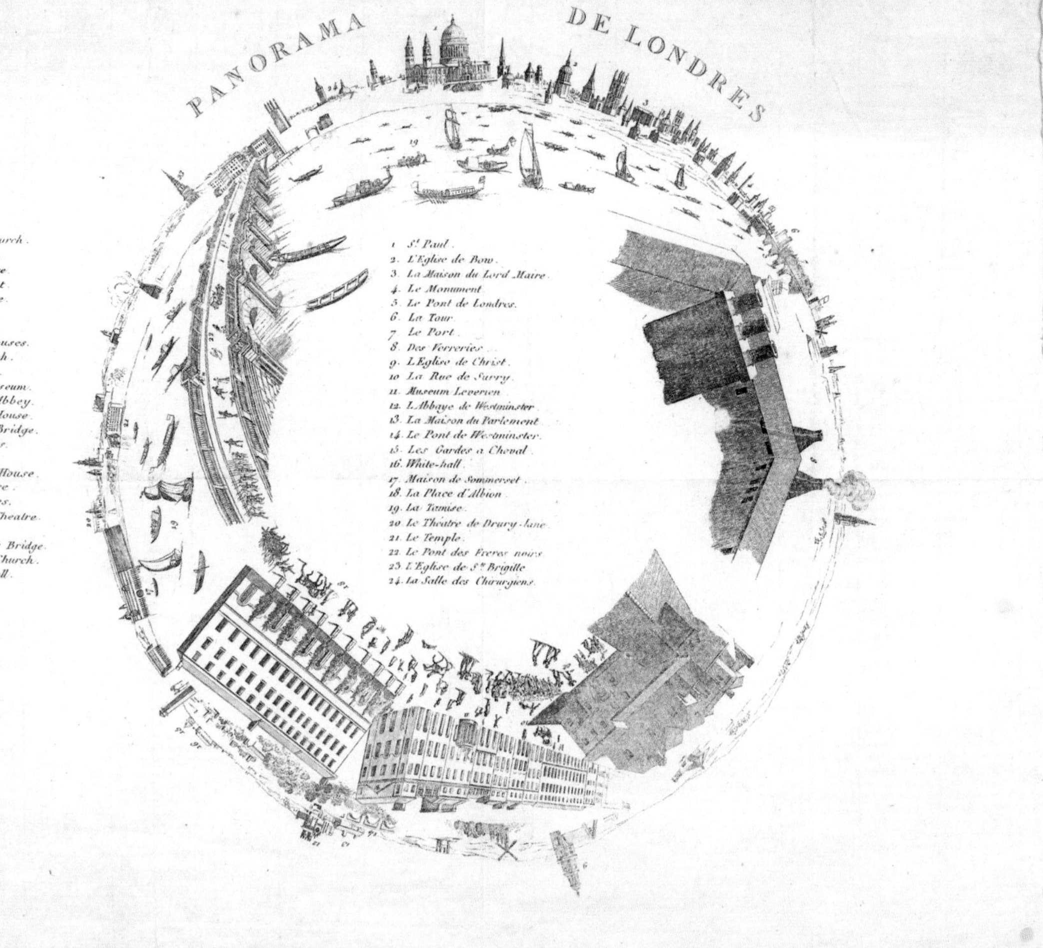

Regency London

74 **CARY, John**

Cary's New and Accurate Plan of London and Westminster.

Publication
London, printed for Jn. Cary, Engraver, and Map-seller No 181 near Norfolk Street, Strand, Corrected to June 1st, 1820.

Description
Large folding engraved map, hand-coloured in outline, dissected and mounted on linen, folding into original green marbled paper slipcase, with publisher's label.

Dimensions
815 by 1528mm (32 by 60.25 inches).

Scale
6½ inches to 1 statute mile.

References
Howgego 184 (17).

A fine example of Cary's map showing London at the end of the Regency Period. The map was extended in 1811 to accommodate the development of the docks on the Isle of Dogs, and Howgego makes particular note of the map:

"The best of the London maps published by the firm was the 'New and Accurate Plan' to a scale of 6½ inches to the mile. First published in 1787, of which at least twenty editions were issued between then and 1825. Unlike many other map-makers Cary was at considerable pains to bring each edition up to date" (Howgego p.21)

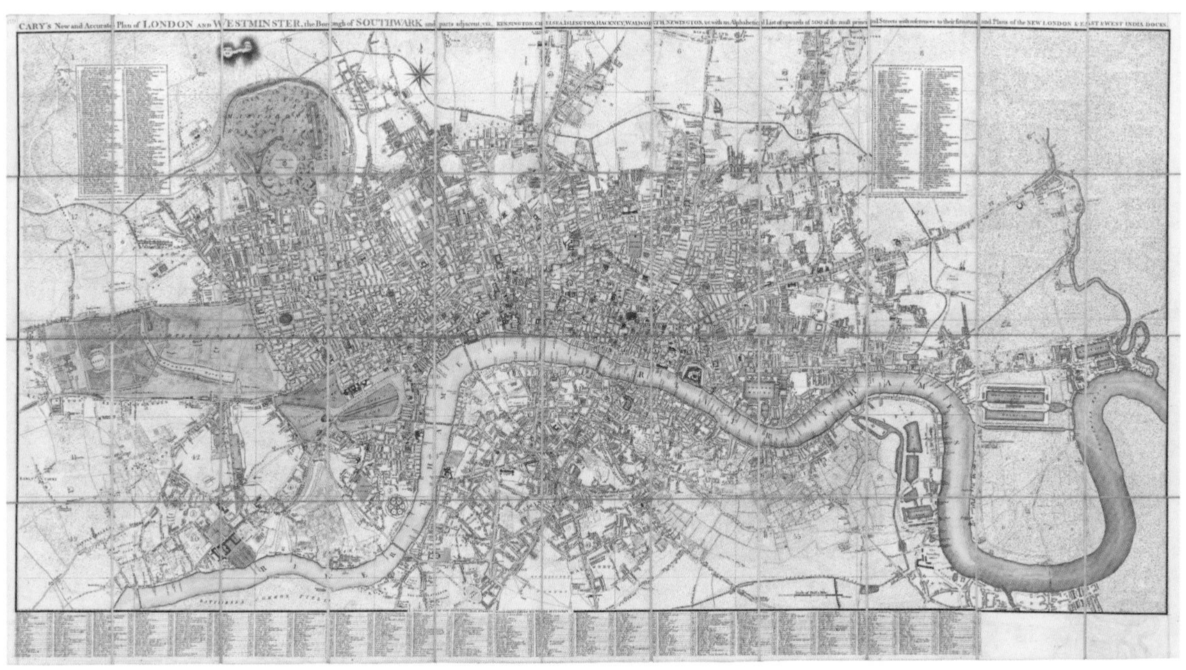

Published under a pseudonym

75 **PERCY, Sholto and Reuben (pseud.)**

A New Map of London, Westminster, Southwark and their Suburbs Drawn from actual Survey, for the History of London, By Sholto & Reuben Percy [pseud.].

Publication
London, Published Octr. 1 1823 by T. Boys, 7 Ludgate Hill, 1823.

Description
Engraved plan, backed on linen, some dampstaining, date on title in manuscript.

Dimensions
255 by 392mm (10 by 15.5 inches).

Scale
2 inches to 1 statute mile.

References
Howgego 294 (1).

This diminutive plan was published in 'The Percy Histories…of all the Capitals of Europe, by Sholto and Reuben Percy', in 1823.

The plan was published under a pseudonym: Reuben Percy was Mr. Thomas Byerley, who was the brother of Sir John Byerley, and the first Editor of 'The Mirror'. Sholto Percy was Mr. Joseph Clinton Robertson, who died in 1852; he was the founder of the 'Mechanics' Magazine'. The name came from Percy Coffee House, in Rathbone Place, where Byerley and Robertson were accustomed to meet from time to time to talk over their joint work.

Hackney fares

76 **PIGOT & Co.**

Pigot & Co.'s New Plan of London.

Publication
London, Published by J. Pigot & Co. 24 Basing Lane, and 16, Fountain Street, Manchester, [c. 1823].

Description
Engraved plan, original hand-colour, some tape marks to old folds, old tears skilfully repaired.

Dimensions
535 by 780mm (21 by 30.75 inches).

Scale
3½ inches to 1 statute mile.

References
Howgego 298 (1).

The plan is coloured to show the boundaries of the City of London, the Cities and "liberties" of Westminster, Southwark, the "rule" of the Kings Bench and Fleet Prison, and the liberties of Southwark. To the upper left of the plan is a detailed explanation: the concentric circles from St Paul's present on the plan are for the calculation of hackney fares, and are said to be no more than one mile apart after "allowing for the angular turns in the streets". Each circle is divided into segments and numbered; the number corresponding to an index of streets and public buildings. The plan shows planned development in Paddington and Bayswater. New developments include Vauxhall Bridge and Millbank Penitentiary; the New London Bridge is shown, but not the approach roads to it and the Thames Tunnel indicated by pecked line.

According to Howgego, the plan appears both in Pigot's 'Metropolitan Guide and Book of Reference', and in Pigot's 'London & Provincial New Commercial Directory for 1824–1825'.

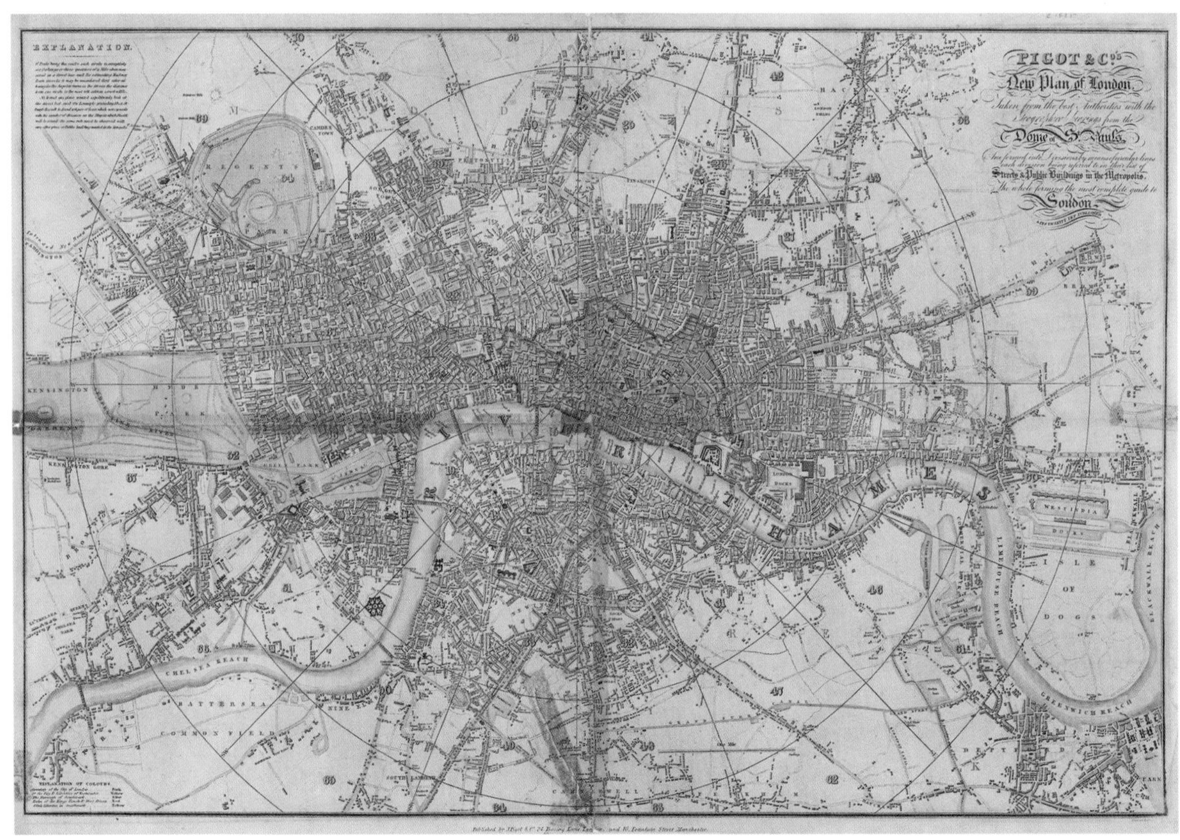

Wyld's plan of London and Westminster

77 **WYLD, James**

A Plan of London and Westminster with the Borough of Southwark. 1825.

Publication
London, Published by James Wyld (Successor to W. Faden) Geographer to the King & to H.R.H. the Duke of York, 3 Charing Cross, 1825.

Description
Engraved plan, original outline hand-colour, dissected and mounted on linen, evenly age-toned, housed within original brown marbled slipcase with publisher's label on front cover, remnants of twentieth century label on rear cover, reading "James Wyld (Successor to Mr. Faden) Geographer to his Majesty and to HRH the Duke of York. London, Charing Cross, opposite Northumberland House", rubbed.

Dimensions
455 by 920mm (18 by 36.25 inches).

Scale
5 inches to 1 statute mile.

References
Howgego 272 (6).

Wyld, the successor to William Faden, continued to issue his London maps, the present example covering the area from Regent's Park in the north west to Bethnal Green in the north east and Chelsea in the south west to Rotherhithe in the south east. To the lower left and lower right of the plan is a list of parishes without the City of London. The boundary of the City of London is marked out in red, the Borough of Southwark in green, the Liberties of Westminster in yellow, and Rules of the Bench and Fleet in orange.

Faden had started issuing these popular and useful maps sometime before 1818; in August 1823 Wyld took over his Charing Cross map business.

CHELSEA REACH
BATTERSEA FIELDS
WESTBOURN GREEN
THE OVAL
Explanation
Public Buildings
Park & Pleasure Grounds
Woods & Plantations
Fields
Gardens
Hills & Rising Grounds
Windmills

MAP of LONDON,
FROM
An Actual Survey made in the Years 1824.1825&182
C. and J. GREENWOOD.
Published by the Proprietors
GREENWOOD, PRINGLE and Co.
13, REGENT STREET, PALL MALL,
London
LONDON DOCK
WEST INDIA
Warehouses
CANAL
LIMEHOUSE
REACH
ISLE OF DOGS
GREENWICH
The Victualling Office
Wells Ship Yard
Bow Common
Majesty
FOURTH
ropolis
Explanation
Boundary of the City of London
Boundary of the City & Liberty of Westminster
Boundary of the Borough of Southwark
Boundary of the Rules of Kings Bench & Fleet Prisons
Boundary of the Clink Liberty
Boundary of Counties
Boundary of Parishes

Greenwood's majestic large-scale plan of London

78 **GREENWOOD, Christopher**

Map of London, from an actual Survey made in the Years 1824.1825 & 1826. By C. and J. Greenwood.

Publication
London, Published, By the Proprietors, Greenwood, Pringle and Co. 13, Regent Street, Pall Mall, August 21st, 1827.

Description
Large engraved map on six sheets, dissected and mounted on linen, fine original outline hand-colour.

Dimensions
1880 by 1290mm (74 by 50.75 inches).

Scale
8 inches to 1 statute mile.

References
Howgego 309 state (1).

Christopher and John Greenwood state in the title that the plan was made from an "Actual Survey", which had taken three years. Plans at the time were often copied from older surveys, or re-issued with minor updating; so conducting a new survey was indeed something to boast about. The plan, which was finely engraved by James and Josiah Neele, is stylistically similar to the Ordnance Survey maps of the time, although it was engraved on a much larger scale of 8 inches to the mile, compared to the OS one inch to the mile. It includes detailed depictions of streets, houses, public buildings, parks, squares, woods, plantations, rivers, hills, windmills; also the marking of the boundaries of the City of London, Westminster, Southwark, Rules of the King's Bench & Fleet Prison, Clink Liberty, counties and parishes. Below the plan is a dedication to George IV, which is flanked by views of Westminster Abbey and St Paul's Cathedral.

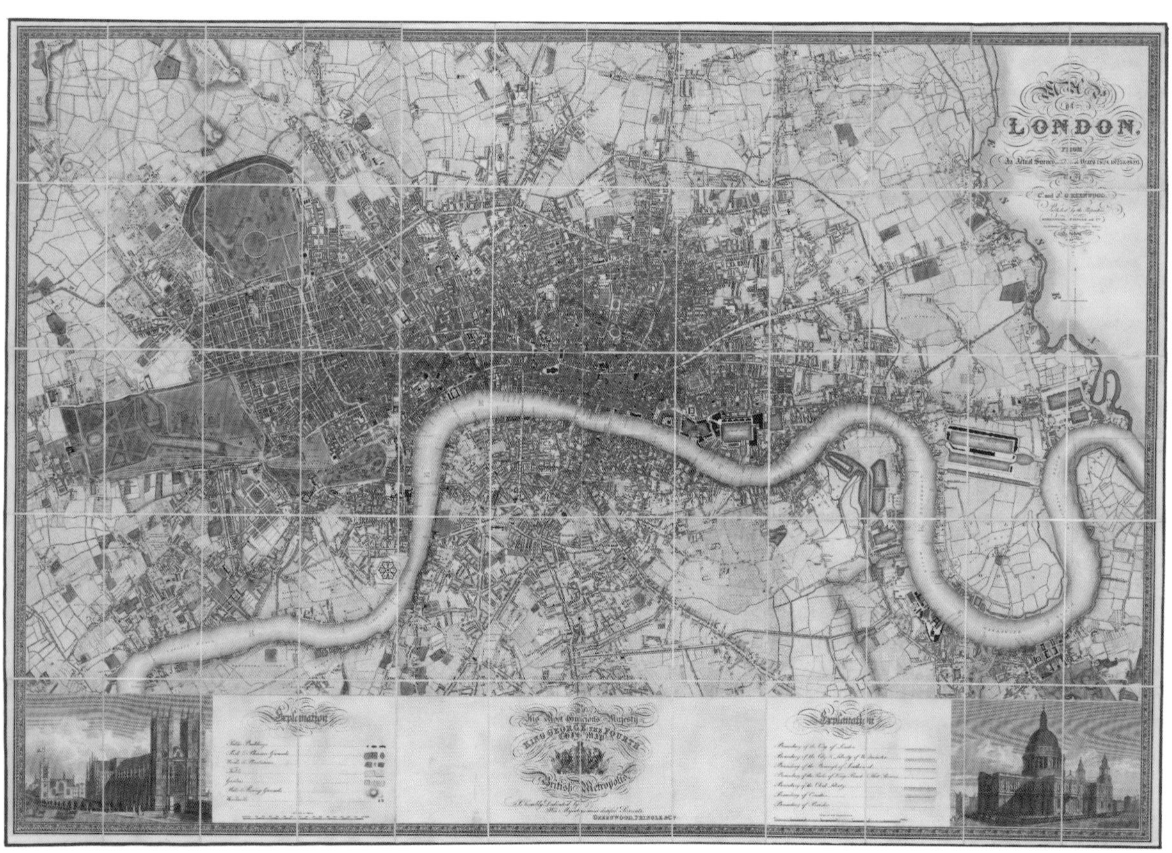

Going south of the river

79 **GREENWOOD, Christopher and John**

Map of London, from an actual Survey made in the Years 1824.1825 & 1826. By C. and J. Greenwood. Extending and Comprising the Various improvements to 1830.

Publication
London, Greenwood & Co., Regent Street, Pall Mall, 31st Augt., 1830.

Description
Large engraved plan on six sheets, fine original hand-colour in outline.

Dimensions
1300 by 1900mm. (51.25 by 74.75 inches).

Scale
8 inches to 1 statute mile.

References
Howgego 309 (2).

The present plan accords with Howgego 309 state (2). The date in imprint is changed to 1830 and the plan is dedicated to King William IV. The map has been extended down another 7¾ inches on the bottom centre sheet (in place of the dedication) to include parts of Stockwell, Camberwell and Peckham. A table of references to parishes has been included on the bottom right-hand sheet and reference numbers have been superimposed on the plan. Additions to the map include the Lea Union Canal, and the names "Bayswater" and "East Greenwich".

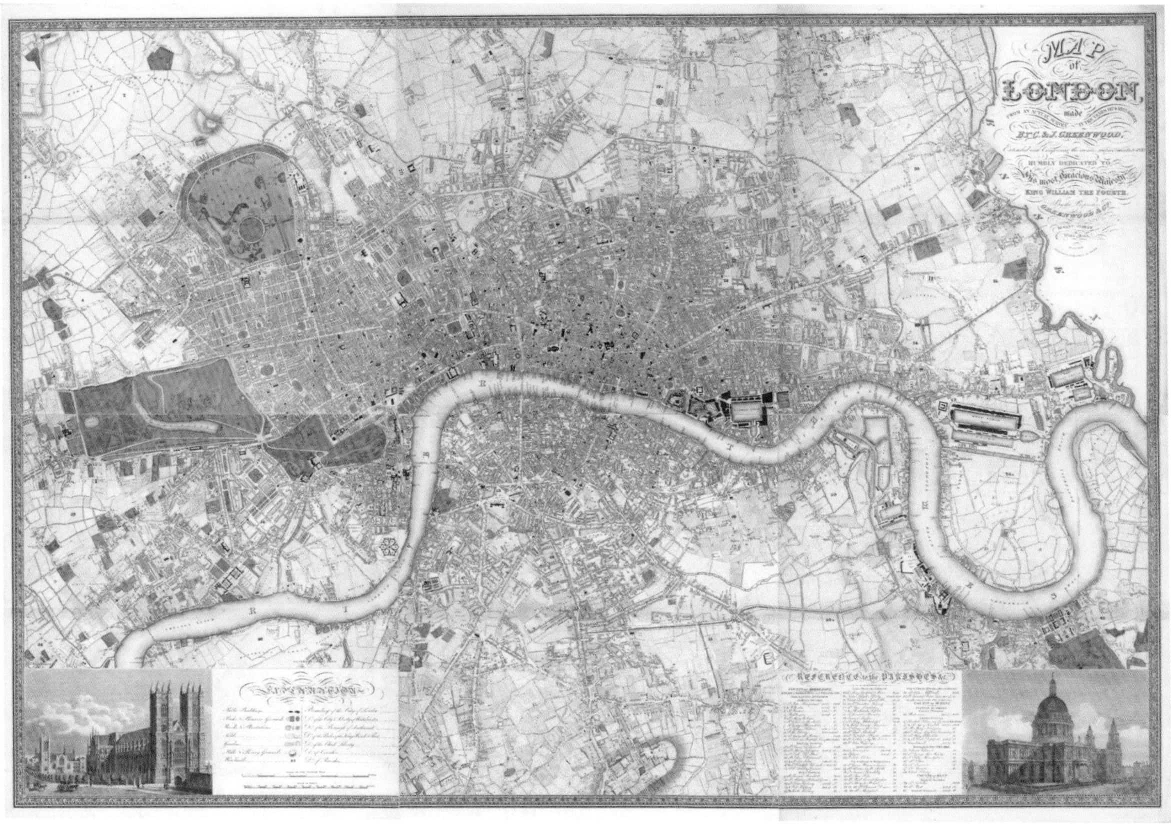

The Great Reform Act

80 **FADEN, William**

A New Topographical Map of the Country in the Vicinity of London, Describing all the New Improvements.

Publication
London, Published by James Wyld (successor to Mr. Faden), Geographer to His Majesty, 5 Charing Cross, July 3, 1829.

Description
Engraved map on two sheets, joined, contemporary hand-colour, marking London's Parliamentary boroughs, extending north to south from Tottenham to Croydon, and west to east from Barnes to Woolwich, tear to upper portion of map skilfully repaired.

Dimensions
930 by 840mm (36.5 by 33 inches).

Scale
2 inches to 1 statute mile.

References
Howgego 250 (4a).

The border is marked off in miles, based on the meridian of St Paul's. The map marks the boundaries of the newly created Parliamentary boroughs of St. Marylebone, Finsbury, Tower Hamlets, and Lambeth – brought into being by the Great Reform Act of 1832. These were in addition to London, the City of Westminster, and Southwark, which had returned members to Parliament for centuries. They bear little relation to modern London constituencies, however, they do reflect the growth of London beyond the traditional areas of the City, Southwark, and Westminster.

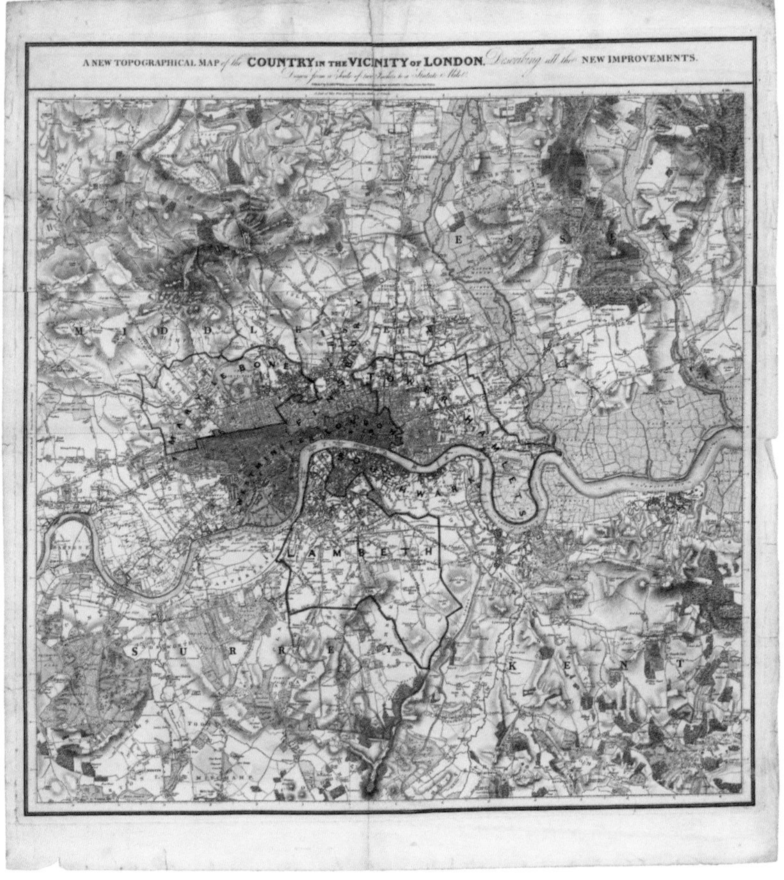

“the intended New Docks of Saint Catherine’s”

81 **CRUCHLEY, [George Frederick]**

Cruchley’s New Plan of London.

Publication
London, Cruchley Map Seller from Arrowsmith’s, 38 Ludgate Street. St. Pauls, [1829].

Description
Engraved plan, dissected and mounted on linen, hand-coloured in outline, three borders engraved with list of street names, housed within original marbled paper slipcase, with publisher’s printed yellow label.

Dimensions
445 by 620mm (17.5 by 24.5 inches).

Scale
$4\frac{7}{8}$ inches to 1 statute mile.

References
Howgego 304 (4)

George Frederick Cruchley (1796–1880) was a publisher of “some of the clearest and most attractive London maps” (Howgego). Cruchley began his cartographic career in the publishing firm of Aaron Arrowsmith. In 1823, he set up on his own at 38 Ludgate Street until 1834, when he moved to 81 Fleet Street. It would appear that he had some help from his former employer, as much of his early output bears in the imprint “From Arrowsmith’s”. He would later acquire a great deal of Arrowsmith’s stock. In 1844, he took over the stock of George and John Cary, which he republished until 1876. In 1877 his entire stock was sold at auction with many of the plates being bought by Gall and Inglis. Cruchley himself died in Brighton in 1880.

This is the first issue of his London map, which was updated to 1846. The 1826 edition showed the London Docks “and the intended New Docks of Saint Catherine’s with all the New Buildings ...” (front cover).

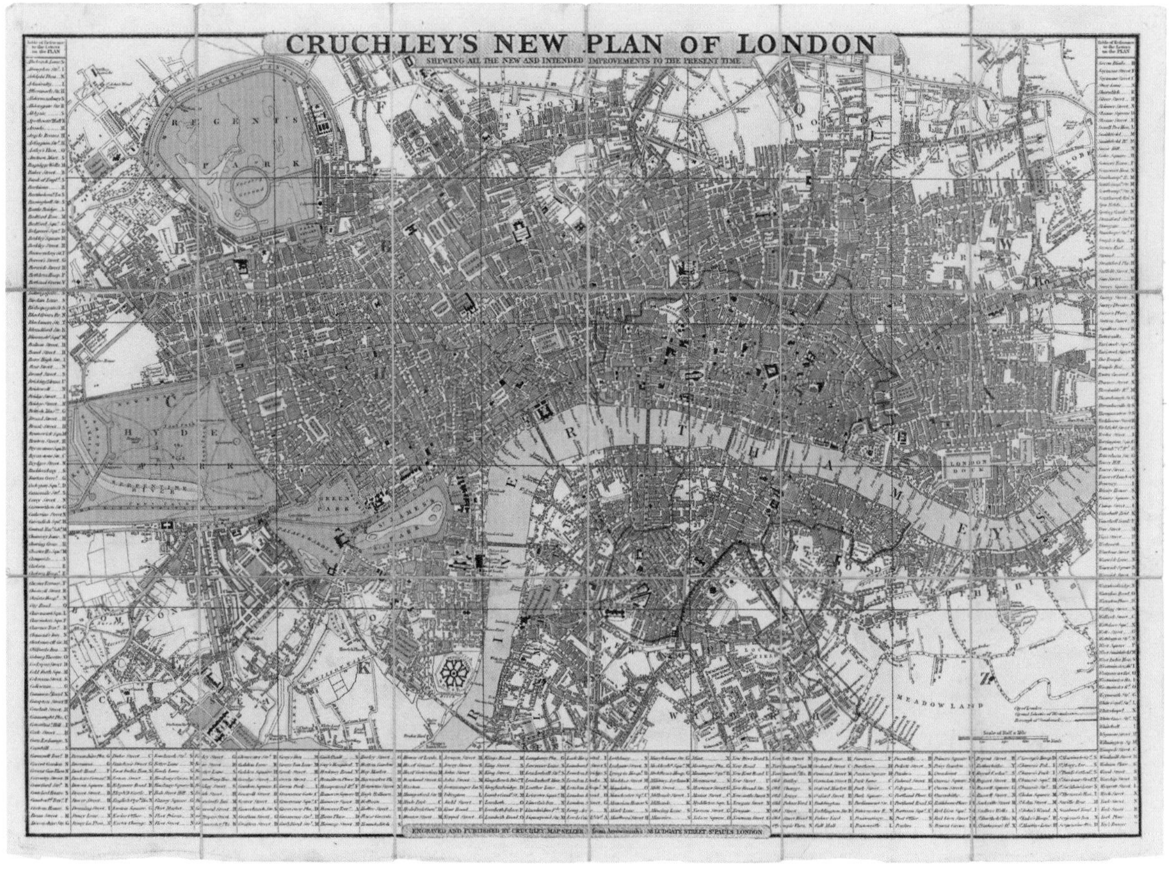

Printed on cloth

82 [Anonymous]

London and its Environs for 1832.

Publication
[London], 1832.

Description
Engraved plan printed in sepia on cloth, a few pale stains.

Dimensions
850 by 980mm (33.5 by 38.5 inches)

Scale
6 inches to 1 statute mile.

References
Howgego 328a (2); this state with date in imprint altered to 1832.

Rare plan of London printed on cloth.

The plan extends west to east from Knightsbridge to Bow, and north to south from Dalston to Camberwell. The plan shows both the Old and New London Bridges. To the upper right is the title with the arms of William IV. The plan is framed by a decorative vine-scroll border with royal arms and garter to the upper centre and allegorical female figures to thc sidcs.

LONDON
and its Environs,
FOR 1832.
LONDON DOCK
BOW COMMON

"a civil but rather a stupid dog"

83 **SHURY, John**

Plan of London From Actual Survey 1833 Presented Gratis to the Readers of the United Kingdom Newspaper by their Obliged Humble Servants The Proprietors.

Publication
London, Engraved and Printed by John Shury, 16 Charterhouse Street, Charterhouse Square, 1833.

Description
Engraved map, old folds and some tears to edges.

Dimensions
562 by 820mm (22.25 by 32.25 inches).

Scale
3¾ inches to one statute mile.

References
Howgego 343 (2).

The map is highly detailed and extends to Islington, Limehouse, Kennington and Brompton. It shows the Surrey Zoological Gardens, opened 1832, New London Bridge and the positions of Lambeth Bridge and Thames Tunnel, then under construction. The ornamental title is embellished with the royal arms and the arms of the City of London and Westminster. The map is bordered on three sides by 33 engraved views of famous buildings including Westminster Abbey, Buckingham Palace and the House of Lords. This is the second state of the map that was originally published in 1832.

John Shury (or Sherry) is mentioned in a letter from Richard Phillips, a distinguished chemist and member of the Royal Society, to Michael Faraday, dated 4 September 1821: "Send to the engraver to wait upon you for the plate – he is a civil but rather a stupid dog – his name is Shury, Charterhouse House [sic] Street, Charterhouse Square." (Letter 148, 'The correspondence of Michael Faraday', by Michael Faraday, Frank A. J. L. James)

Title in panel across the top, in the center the royal arms together with those of Westminster and Southwark. At bottom right scale-bar and explanation of boundary symbols. Greek key border. Engraver/printer details in margin at bottom left continuing on bottom right.

84 **SHURY, John**

Plan of London From Actual Survey 1839 Published by Orlando Hodgson, 111, Fleet Street London.

Publication
London, Published by Orlando Hodgson, 111 Fleet Street, 1839.

Description
Engraved map, old folds backed with japan paper.

Dimensions
520 by 860mm (20.5 by 33.75 inches).

Scale
3¾ inches to one statute mile.

References
Howgego 343 (intermediate state between 4 & 5).

Publisher's details in margin at bottom left continuing on bottom right. In left, right and bottom margins are engravings of 33 London landmarks. The engraving of the Colosseum has been re-sited at lower left. The map shows the London & Greenwich Railway, the London & Birmingham Railway, and the Hungerford Suspension Bridge.

The present map is an intermediate state between Howgego 343 states (4) and (5). The map bears the imprint of Orlando Hodgson as in the fourth state, but with the date in the title changed to 1839.

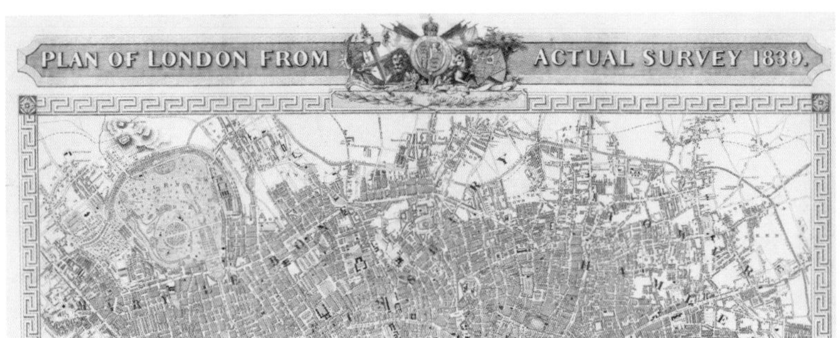

Item 84 (detail)

---

85 **SHURY, John**

Plan of London from Actual Survey 1840.

Publication
[London], 1840.

Description
Engraved map, old folds backed with japan paper.

Dimensions
578 by 818mm (22.75 by 32.25 inches).

Scale
3¾ inches to one statute mile.

References
Howgego 343 (intermediate state between 4 & 5).

To the upper left and right are vignette views of Greenwich Hospital and London from Waterloo Bridge respectively. At the bottom right is a scale-bar and explanation of boundary symbols. The plan is framed by a Greek key border, and 29 engraved views of London landmarks, including: St Pauls, Temple Bar, The Colosseum, St. James's Place, and Somerset House.

The map was first issued as a free supplement with the United Kingdom Newspaper in 1832.

Item 85 (detail)

"A complete bijou of a map, not less useful than elegant"

86 **FRASER, James**

James Fraser's Panoramic Plan of London.

Publication
London, 215 Regent Street, James Fraser, 1837.

Description
Engraved map, dissected and mounted on linen, hand-coloured in outline, bordered with 18 engraved views of some of the principal attractions of London, housed within original green cloth folder, bookseller's ticket on pastedown "Sold by John Bubb, Bradford", with printed label on upper cover, spine worn and repaired.

Dimensions
410 by 550mm (16.25 by 21.75 inches).

References
Howgego 332 (3).

The eighteen views surrounding the map depict the New Post Office, Italian Opera House, Queen's New Palace, Entrance to Hyde Park, Custom House, National Scottish Church, Drury Lane, Somerset House, Westminster Abbey, The Colosseum, Covent Garden Theatre, Thames Tunnel, St Paul's, Waterloo Bridge, Bank of England, New London Bridge, London University and Hammersmith Suspension Bridge.

According to Fraser's advertisement, the map was available at four different prices, starting at 7s 6d in a French Case to 16s in a black frame; our example is the more humble 3s issue in cloth. The reviews were handsome, the Literary Gazette calling the publication "A complete bijou of a map, not less useful than elegant." The Athenaeum regarded the map as "generally serviceable", leaving the Christian Remembrancer to throw caution to the wind and claim "the eighteen marginal views of the principle public buildings are perfect gems of art, and cannot fail, when combined with the correct delineation of the survey, to recommend it to the public favour both for elegance and unity."

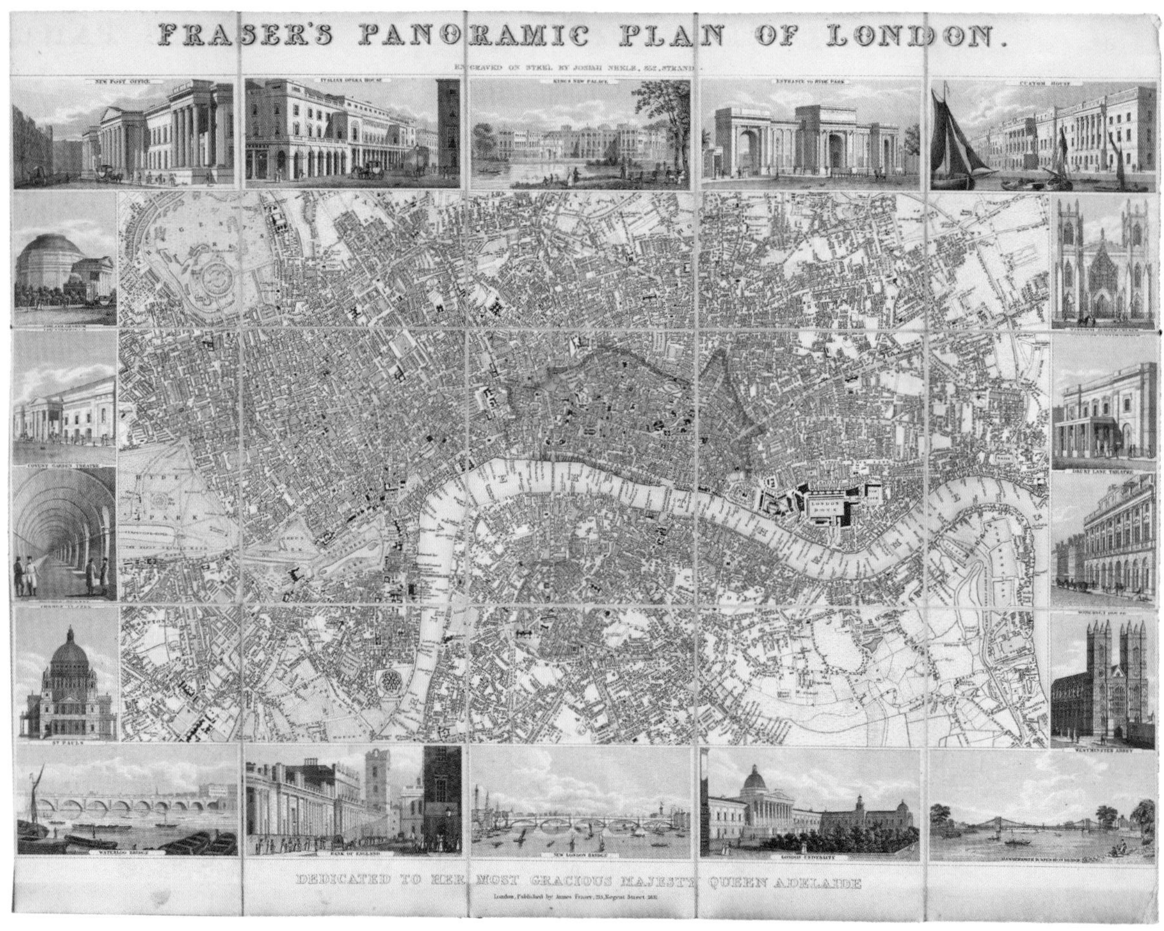

The first steam railway to have a terminus in the capital

87 **WALLIS, Edward**

Wallis's Guide for Strangers through London.

Publication
[London], Published by E. Wallis, 42, Skinner Street, [c.1840].

Description
Engraved plan, fine original hand-colour, alphabetical list of streets below, old folds reinforced, some with minor loss.

Dimensions
540 by 750mm (21.25 by 29.5 inches).

Scale
6 inches to one mile.

References
Howgego 301 (intermediate state between 8 and 9).

Wallis's Guide for Strangers was a popular map, printed in numerous editions between 1813 and 1843. This edition shows the London and Greenwich Railway, opened between 1836 and 1838, which terminated at London Bridge, and was the first steam railway to have a terminus in the capital. Also shown is the London and Blackwall line, opened on July 6, 1840. This terminates at Minories on the current map, but was extended to Fenchurch Street in 1841. The London and Birmingham Railway was the first intercity line to be built into London. It terminates at Euston Station, which on this map is named as the Birmingham Railway Depot, and ran from 1833 to 1846, when it became part of the London and North Western Railway Line. Lastly is the Eastern Counties Railway, opened June 20, 1839. In July of 1840, it was extended to run from Shoreditch, as shown on this map, to Brentwood.

In addition to selling maps and charts, Edward Wallis, with his father John, was the most prolific publisher of educational board games. His father was the inventor of the jigsaw puzzle.

Although not dated, the plan can be dated to 1840 due to the Eastern Counties Railway to Shoreditch (1840), and London & Blackwall to Minories (1840). This would place it between state (8) and (9) in Howgego.

Bauerkeller's rare and strikingly modern embossed plan of London

88 **BAUERKELLER, G[eorge]**

Bauerkeller's New Embossed Plan of London 1841.

Publication
Paris & London, Published by Ackerman & Co., 96 Strand, 1841.

Description
Hand-coloured engraved embossed map, dissected and mounted on linen, inset of the Isle of Dogs, tables of the population of London and key to the map's colouring, a few minor tears to margins, skilfully repaired, folding into brown paper slipcase, with tree branch pattern.

Dimensions
680 by 1160mm (26.75 by 45.75 inches).

Scale
6 inches to 1 statute mile.

References
Howgego 377.

This extraordinary embossed plan shows each locality in a different colour and built up areas raised in white, similar to the technique of Braille. Extending from Islington in the north to Kennington in the south and from Kensington High Street in the west to the West India Docks, with an inset of Greenwich at a smaller scale. A unique style of cartography in very good condition.

In an advertisement in the 'Sporting Magazine Advertiser' Ackermann announces its publication: "... The Buildings are raised, and,

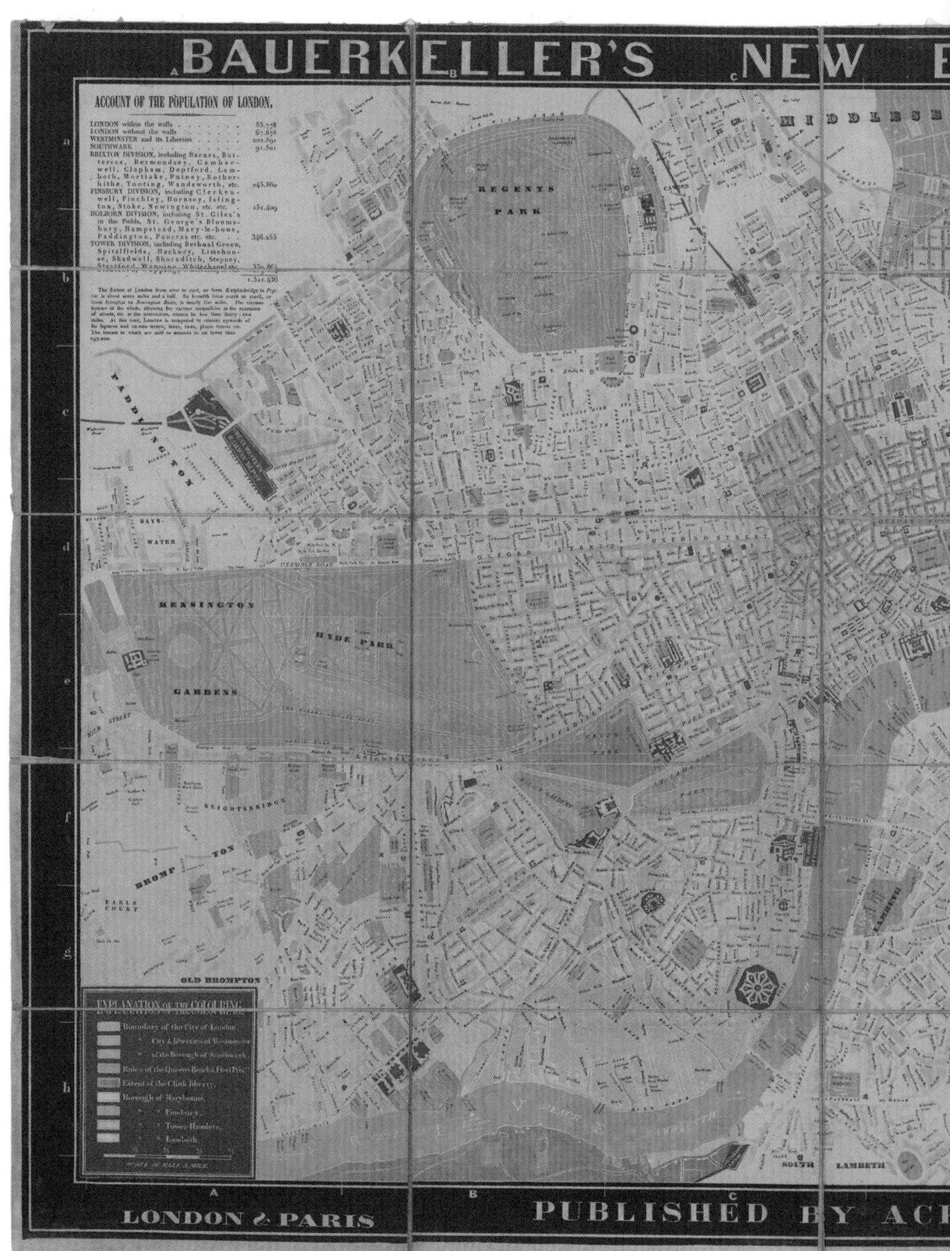

with the Railroads, Parks, Squares, &c. apper very prominent. The Parishes are also distinguished in delicate tints, and the entire arrangement is so remarkably conspicuous that, whether for the Visitor or the Office, its utility will be generally acknowledged" (Howgego).

A table to the upper left of the plan records a population of just over 1.5 million in 1841, with 122,000 residing in the City of London. The number of houses is estimated to be above 197,000; there are over 80 squares and some 10,000 streets.

A fine and detailed plan of early Victorian London

89 **CRUCHLEY, G[eorge] F[rederick]**

Cruchley's New Plan of London improved to 1844.

Publication
London, Engraved & Published By G.F. Cruchley, No.81 Fleet Street, 1844.

Description
Hand-coloured engraved plan, dissected and mounted on linen, with title and border pasted down, edged in green silk, publisher's advertisements to verso, folding into original green cloth slipcase with publisher's label, rubbed and scuffed.

Dimensions
560 by 1430mm (22 by 56.25 inches).

Scale
4⅞ inches to 1 statute mile.

References
Howgego 304 C (7).

George Frederick Cruchley (1796–1880) was a publisher of "some of the clearest and most attractive London maps" (Howgego). The present map, published on a scale of almost five inches to the mile, was first issued in 1826, with Cruchley re-issuing it several times for the following 20 years, each time updating it with the latest additions.

Cruchley began his cartographic career in the publishing firm of Aaron Arrowsmith. In 1823, he set up on his own at 38 Ludgate Street until 1834, when he moved to 81 Fleet Street. It would appear that he had some help from his former employer, as much of his early output bears in the imprint "From Arrowsmith's". He would later acquire a great deal of Arrowsmith's stock. In 1844, he took over the stock of George and John Cary, which he republished until 1876. In 1877 his entire stock was sold at auction with many of the plates being brought by Gall and Inglis. Cruchley himself passed away in Brighton in 1880.

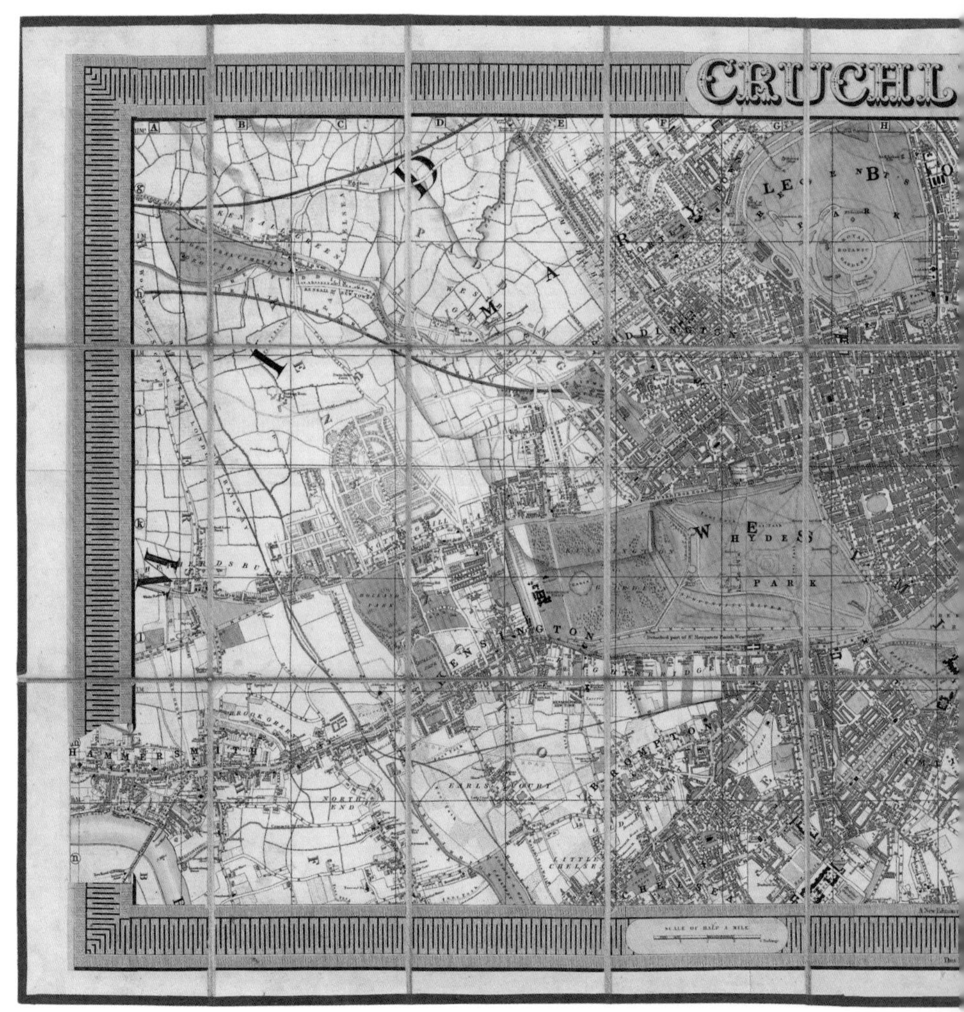

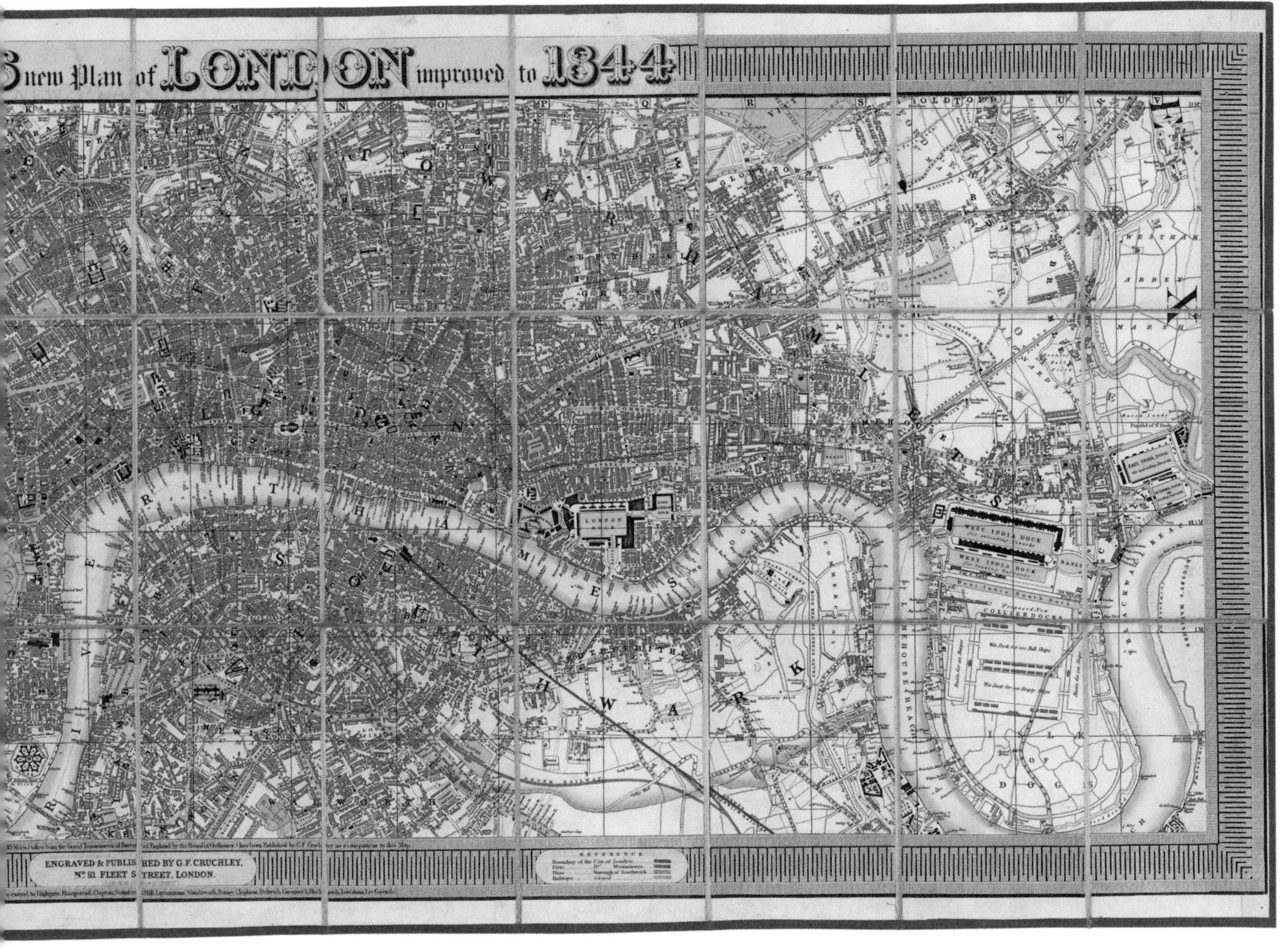
new Plan of LONDON improved to 1844
ENGRAVED & PUBLISHED BY G. F. CRUCHLEY,
No. 31 FLEET STREET, LONDON.
WEST INDIA DOCK
ISLE OF DOGS
LONDON DOCKS

Docklands

90 **[Anonymous]**

Plan of the London Docks, 1851.

Publication
London, Lith. Waterlow & Sons, 65 to 68 London Wall, 1851.

Description
Hand-coloured lithograph plan, note to plan lower left, view of the London Docks lower right.

Dimensions
595 by 920mm (23.5 by 36.25 inches).

Scale
Approx. 42 inches to 1 statute mile.

The London docks, situated in Wapping, were built between 1799 and 1815 at a cost exceeding £5.5 million. London by this point had outgrown the wharves on the Thames, and more capacity was desperately needed. The principal architects were Daniel Alexander and John Rennie. The docks specialised in high-value items such as ivory, spices, coffee, cocoa, wine and wool. The docks would later be amalgamated with St. Katherine Docks.

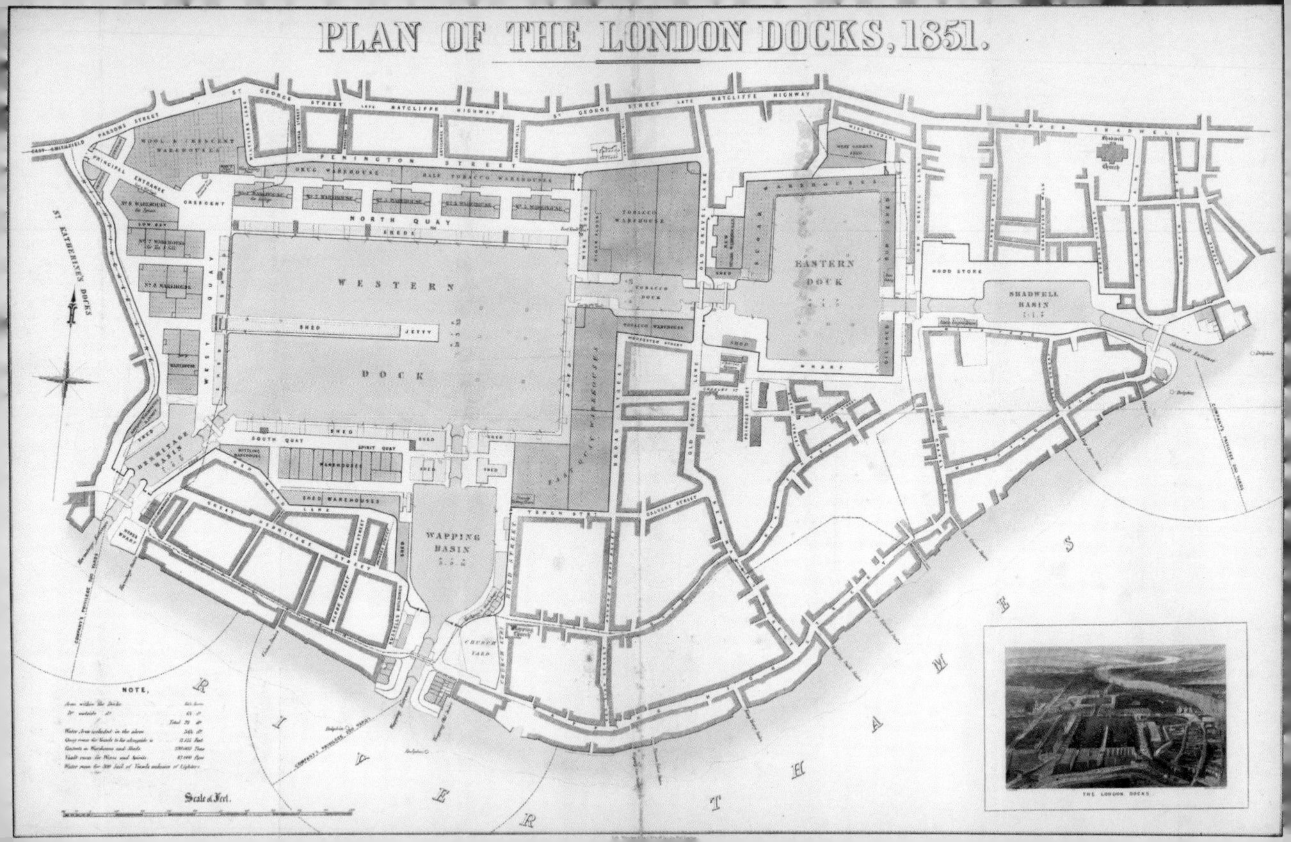
PLAN OF THE LONDON DOCKS, 1851.
ST GEORGE STREET LATE RATCLIFFE HIGHWAY
PENINGTON STREET
UPPER SHADWELL
WOOL & CRESCENT WAREHOUSES
DRUG WAREHOUSE
HALF TOBACCO WAREHOUSES
PRINCIPAL ENTRANCE
CRESCENT
NORTH QUAY
WEST QUAY
SOUTH QUAY
SPIRIT QUAY
WESTERN DOCK
SHED
JETTY
TOBACCO WAREHOUSE
TOBACCO DOCK
SUGAR WAREHOUSES
EASTERN DOCK
WOOD STORE
SHADWELL BASIN
OLD GRAVEL LANE
NEW GRAVEL LANE
WORCESTER STREET
BROAD STREET
EAST QUAY WAREHOUSES
BOTTLING WAREHOUSE
WAREHOUSES
SHED WAREHOUSES
HERMITAGE BASIN
ST KATHERINE'S DOCKS
GREAT HERMITAGE STREET
TENCH STREET
WAPPING BASIN
CHURCH YARD
RIVER THAMES
NOTE,
Scale of Feet.
THE LONDON DOCKS

The Great Exhibition

91 **REYNOLDS, James**

Reynolds's Map of London with the Latest Improvements.

Publication
London, J. Reynolds, 174 Strand, 1851.

Description
Hand-coloured engraved folding map, 'Drawn and Engraved by H. Martin, 8 Dyers Buildings, Holborn'; a few old tears but generally an attractive map, together with the booklet of 22, [14] with an inserted outline map of "exibition Chief Objects of Interest in London"; folding into original cloth boards, the upper cover with a green printed label 'The Exhibition Map of London and Visitors Guide. price 1s 6d Colored.'

Dimensions
435 by 770mm (17.25 by 30.25 inches).

Scale
4 inches to 1 statute mile.

References
Howgego 406 (3).

An attractive map and guide to London produced for visitors during the Great Exhibition.

The plan shows the Crystal Palace, which occupies a large part of the southern portion of Hyde Park. Railways are clearly marked in black with main roads coloured yellow, open spaces in green, and the City of London outlined in red. A booklet accompanies the map, which contains a great deal of information for the "Stranger" to London. It advises him that he must be "vigilant and circumspect… lest he become the prey of some of the swarms of knaves". After frightening the poor tourist half to death the guide goes on to list the various amusements and sights the city has to offer. Most notable among its attractions is the Great Exhibition, which if one wishes to go to the opening day will cost a gentleman £3 3s and a lady £2 2s.

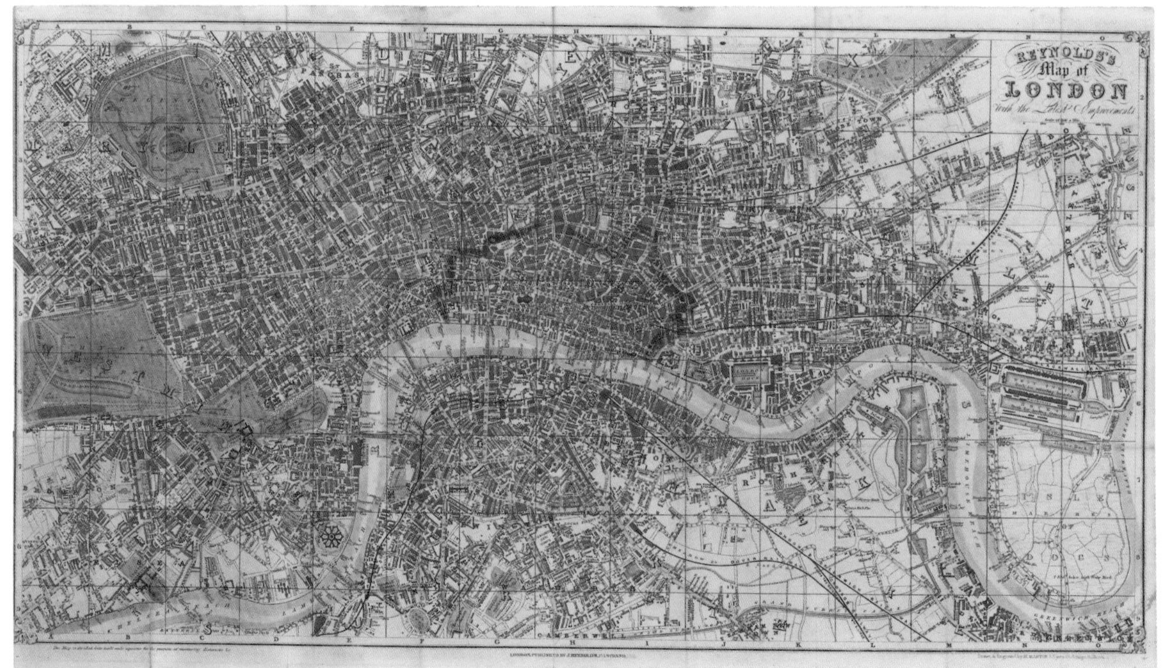

"Wyld's Model of the Earth"

92 **WYLD, James**

Wyld's New Plan of London. Nouveau Plan de Londres. Neuer Plan de London.

Publication
London, Published by James Wyld Geographer to the Queen and H.R.H. Prince Albert Charing Cross East (Opposite Northumberland Street), [c.1852].

Description
Wood-engraved map, original hand-colour, title repeated in French and German, eleven vignette views of London landmarks, key repeated in French and German.

Dimensions
555 by 950mm (21.75 by 37.5 inches).

Scale
3 5/8 inches to 1 statute mile.

References
Hyde 25 (4).

James Wyld published this map in the year of the Great Exhibition, 1851. The exhibition hall, or Crystal Palace, is picked out in red, and occupies a large proportion of the south side of Hyde Park. The Crystal Palace is also depicted in one of the views to the map's border, its vast scale stretching out into the distance. Also depicted in the border is "Wyld's Model of the Earth in Leicester Square". The globe, which measured 18.39m in diameter, was housed within a purpose-built hall in the middle of Leicester Square, and was open to the general public from 1851 to 1862. The globe itself consisted of an internal staircase with a representation of the world depicted in plaster-of-paris on its interior surface. In the surrounding galleries were displays of Wyld's maps, globes and surveying equipment.

The present map would appear to be issued in 1852 or later, as the Great Northern Railway is shown to Kings Cross.

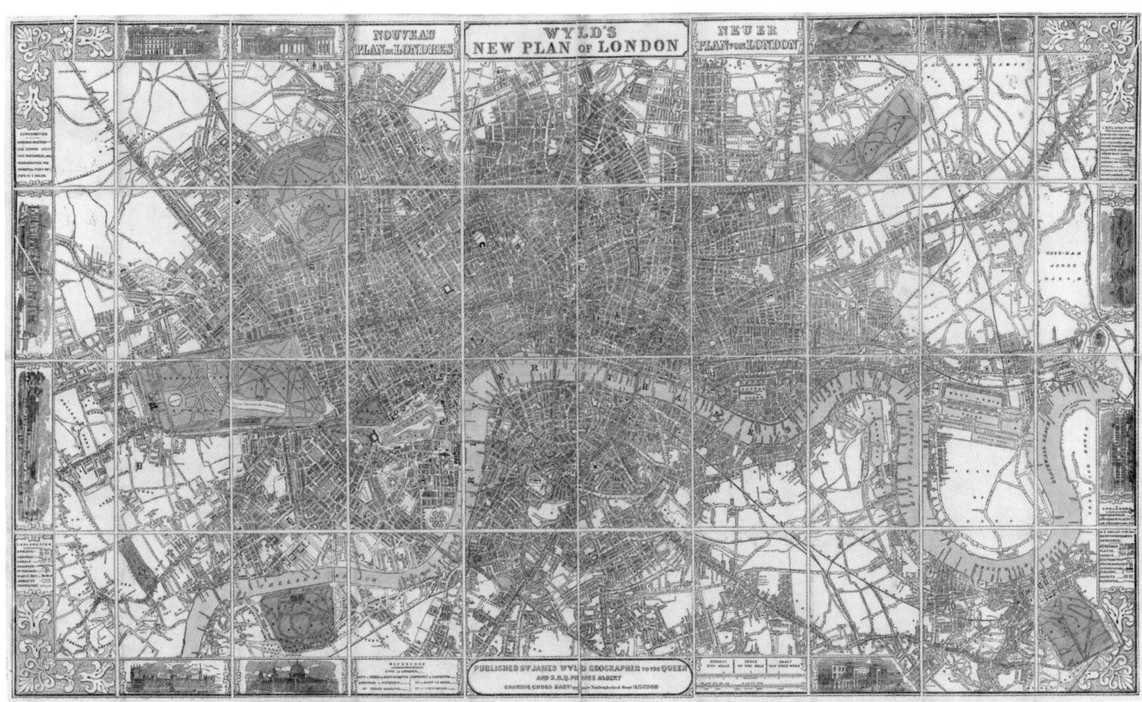

A birds-eye view

93 **WHITTOCK, Nathaniel and Edmund WALKER**

London in the Reign of Queen Victoria.

Publication
London, Lloyd Brothers & Co., 96 Gracechurch Street, January 1st, 1859.

Description
Long tinted lithographic panorama marked as Proof in the stone, wide margins; restorations to folds.

Dimensions
1200 by 340mm (47.25 by 13.5 inches).

Scale
4⅞ inches to 1 statute mile.

Proofs, early impressions on better paper, were priced at £2 2s, double the price of an ordinary print. Nathaniel Whittock made the drawings for this panoramic birds-eye view as from above Southwark. "Though by what means could in that locality have attained such an elevation that enabled him to draw his plan,we are at a loss to conceive" marvelled a contemporary writer in the Art Journal (vol. V, p. 128). Edmund Walker transferred Whittock's drawing onto stone and the leading lithographers of the day printed it.

REIGN OF QUEEN VICTORIA

Taxi!

94 [Anonymous]

The Royal Courts of Justice Central Hotel. Limited. Sketch Map Showing its Situation by a Shilling Cab Fare Radius.

Publication
London, Jas. Truscott & Son, Lith., [c.1882].

Description
Lithograph 2 ff. pamphlet, with map, two elevations of a proposed hotel, and a ground floor plan, a few nicks and tears to old folds.

Dimensions
470 by 600mm (18.5 by 23.5 inches).

The plan shows the shilling cab radius, which was at the time 2 miles from Charing Cross. The plan can be reasonably accurately dated by the marking of the Royal Courts of Justice, as the "New Law Courts", which were opened by Queen Victoria in 1882.

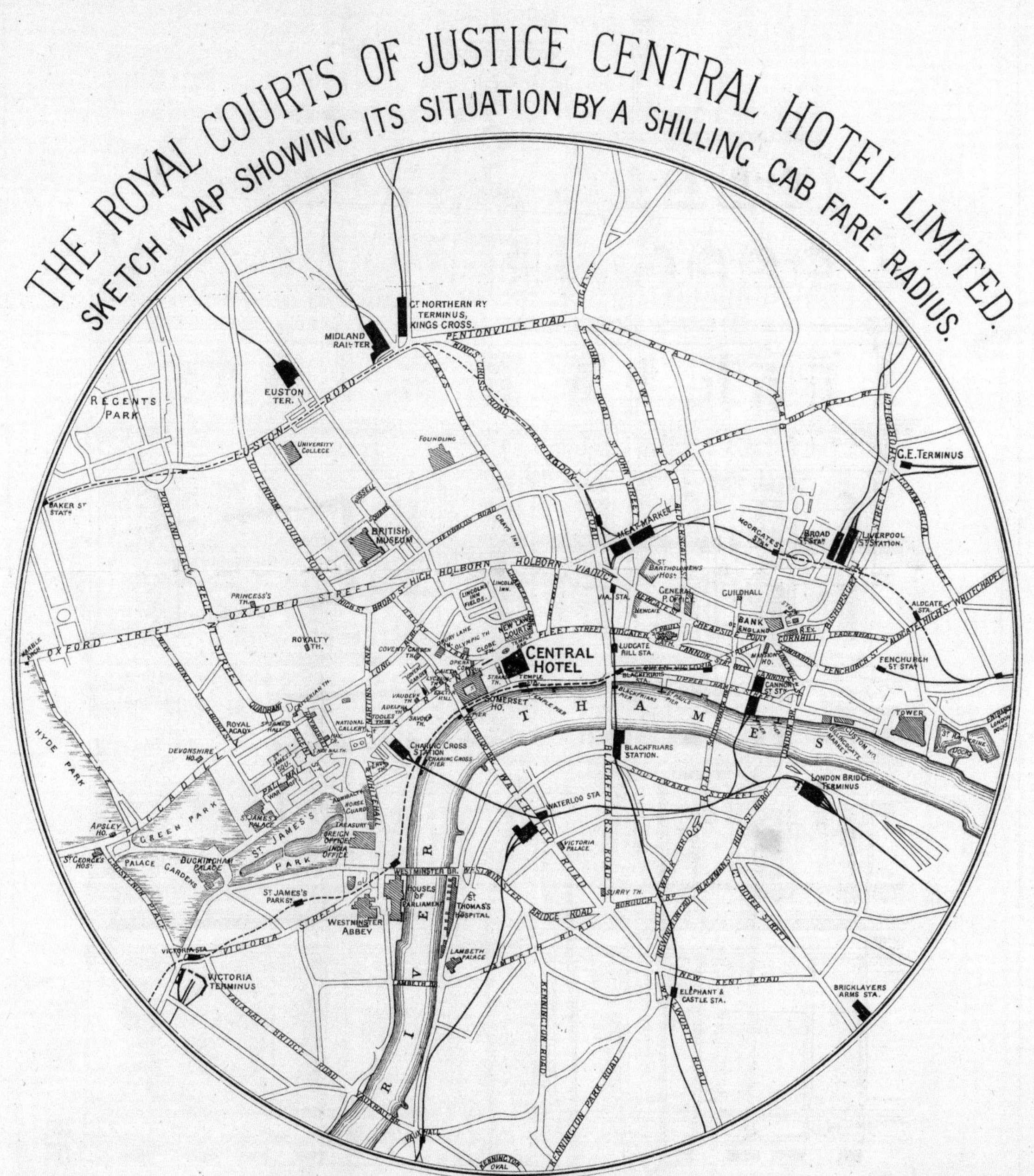

THE ROYAL COURTS OF JUSTICE CENTRAL HOTEL. LIMITED.
SKETCH MAP SHOWING ITS SITUATION BY A SHILLING CAB FARE RADIUS.
CENTRAL HOTEL
GT. NORTHERN RY TERMINUS, KINGS CROSS.
MIDLAND RAILY TER.
EUSTON TER.
REGENTS PARK
PENTONVILLE ROAD
CITY ROAD
EUSTON ROAD
UNIVERSITY COLLEGE
FOUNDLING
BAKER ST STATN
BRITISH MUSEUM
RUSSELL SQUARE
TOTTENHAM COURT ROAD
PORTLAND PLACE
REGENT STREET
OXFORD STREET
HIGH HOLBORN
HOLBORN VIADUCT
PRINCESS'S TH.
ROYALTY TH.
MARBLE ARCH
LINCOLNS INN FIELDS
LINCOLNS INN
NEW LAW COURTS
FLEET STREET
LUDGATE HILL STA.
MEAT MARKET
ST BARTHOLOMEWS HOSP
GENERAL POST OFFICE
NEWGATE
GUILDHALL
BANK OF ENGLAND
CHEAPSIDE
CORNHILL
MANSION HO.
MOORGATE ST STA.
BROAD ST STATN
LIVERPOOL ST STATION.
G.E. TERMINUS
SHOREDITCH
COMMERCIAL ST
WHITECHAPEL
ALDGATE STA.
FENCHURCH ST STA.
CANNON ST STN
BLACKFRIARS STA.
TOWER
CUSTOM HO.
ST KATHERINE DOCKS
SOMERSET HO.
TEMPLE STA.
CHARING CROSS STATION
NATIONAL GALLERY
ROYAL ACADY
ST JAMES'S HALL
DEVONSHIRE HO.
HYDE PARK
GREEN PARK
APSLEY HO.
PICCADILLY
ST JAMES'S PALACE
ST JAMES'S PARK
BUCKINGHAM PALACE
PALACE GARDENS
HORSE GUARDS
ADMIRALTY
TREASURY
FOREIGN OFFICE
INDIA OFFICE
WHITEHALL
ST GEORGES HOSP.
GROSVENOR PLACE
WESTMINSTER BR.
HOUSES OF PARLIAMENT
WESTMINSTER ABBEY
ST JAMES'S PARK ST.
ST THOMAS'S HOSPITAL
VICTORIA STREET
VICTORIA STA.
VICTORIA TERMINUS
VAUXHALL BRIDGE ROAD
LAMBETH PALACE
LAMBETH BR.
RIVER THAMES
WATERLOO STA
VICTORIA PALACE
SURRY TH.
BLACKFRIARS ROAD
BLACKFRIARS STATION.
SOUTHWARK STREET
SOUTHWARK BRIDGE
BOROUGH RD
LONDON BRIDGE TERMINUS
DOVER STREET
NEW KENT ROAD
ELEPHANT & CASTLE STA.
BRICKLAYERS ARMS STA.
KENNINGTON ROAD
KENNINGTON PARK ROAD
KENNINGTON OVAL
VAUXHALL
WALWORTH ROAD
JAS TRUSCOTT & SON, LITH. LONDON.

"The Lowest Class. Vicious, semi-criminal"

95 **BOOTH, Charles**

Descriptive Map of London Poverty 1889.

Publication
London, 1889.

Description
Lithograph plan on four sheets, printed in colours, key to plan below each sheet.

Dimensions
1050 by 1250mm (41.25 by 49.25 inches).

Scale
6 inches to 1 statute mile.

References
Hyde 252.

"Quite the most important thematic maps of the Metropolis in the nineteenth century were those which accompanied Charles Booth's Monumental survey." (Hyde)

A fascinating map of fundamental importance to British social reform. Based upon Stanford's 'Library Map of London'. The colouring of the map depicts, by street: "The Lowest Class. Vicious, semi-criminal" (black); "Very Poor, casual. Chronic Want" (blue); "Poor. 18s to 21s a week for a moderate family" (light blue); "Mixed. Some comfortable, others poor" (purple); "Fairly Comfortable. Good ordinary earnings" (pink); "Well-to-do. Middle class" (red); "Upper-middle and Upper classes. Wealthy" (yellow).

Charles Booth (1840–1916), shipowner and writer on social questions, began his long and successful career as a shipowner at the age of twenty-two, when he joined his eldest brother Alfred as partner in Alfred Booth & Co. He grew up with the Trade Union movement, and in general sympathy with its earlier policy, but its later developments he regarded with misgiving.

Booth had always taken an interest in the welfare of working men, but it was not until he was past middle age that there began to appear the works which established his reputation as a writer on social questions, including his "inquiry into the condition and occupations of the people of London", the earlier part of which appeared, along with this map, as Labour and Life of the People (1889), and the whole as Life and Labour of the People in London (1891–1903). Booth's works appeared at a critical time in the history of English social reform. A lively interest was being taken in the problems of pauperism, and it was coming to be recognized that benevolence, to be effective, must be scientific. Life and Labour was designed to show "the numerical relation which poverty, misery and depravity bear to regular earnings and comparative comfort, and to describe the general conditions under which each class lives". Among the many who helped him to compile his material, and edit it, were his wife's cousin, Miss Beatrice Potter (Mrs. Sidney Webb) and (Sir) Graham Balfour for the earlier volumes, and Ernest Aves for the later. It was no proper part of Booth's plan to analyse economic changes or to trace the course of social development. His object was to give an accurate picture of the condition of London as it was in the last decade of the nineteenth century. In this light, his 'Life and Labour' was recognized as perhaps the most comprehensive and illuminating work of descriptive statistics which had yet appeared.

Booth married in 1871 Mary, only daughter of Charles Zachary Macaulay, and granddaughter of Zachary Macaulay. There were three sons and four daughters of the marriage. He died 23 November 1916 at his home, Gracedieu Manor, Whitwick, and was buried at Thringstone, Leicestershire.

DESCRIPTIVE MAP OF LONDON POVERTY 1889.

North-Western sheet, comprising part of Hampstead; Paddington (excepting north-west corner); Parts of St. George's Hanover Square, Westminster, Strand, Holborn and Islington; the whole of St. Giles's and Marylebone; and most of St. Pancras.

DESCRIPTIVE MAP OF LONDON POVERTY 1889.

North-Eastern sheet, comprising parts of Hackney, Islington, and Holborn; the whole of the City, Shoreditch, Bethnal Green, Whitechapel, St. George's-in-the-East, Stepney, Mile End, and Part of Poplar.

THE STREETS ARE COLOURED ACCORDING TO THE GENERAL CONDITION OF THE INHABITANTS, AS UNDER:—

- Lowest class. Vicious, semi-criminal.
- Very poor, casual. Chronic want.
- Poor. 18s. to 21s. a week for a moderate family.
- Mixed. Some comfortable, others poor.
- Fairly comfortable. Good ordinary earnings.
- Well-to-do. Middle class.
- Upper-middle and Upper classes. Wealthy.

A combination of colours—as dark blue and black, or pink and red—indicates that the street contains a fair proportion of each of the classes represented by the respective colours.

THE STREETS ARE COLOURED ACCORDING TO THE GENERAL CONDITION OF THE INHABITANTS, AS UNDER:—

- Lowest class. Vicious, semi-criminal.
- Very poor, casual. Chronic want.
- Poor. 18s. to 21s. a week for a moderate family.
- Mixed. Some comfortable, others poor.
- Fairly comfortable. Good ordinary earnings.
- Well-to-do. Middle class.
- Upper-middle and Upper classes. Wealthy.

A combination of colours—as dark blue and black, or pink and red—indicates that the street contains a fair proportion of each of the classes represented by the respective colours.

One of the most important maps of Victorian London

96 **STANFORD, Edward**

Library Map of London and its Suburbs.

Publication
London, Edward Stanford, 1891.

Description
Engraved map, hand coloured in outline on four sheets, each segmented into 25 segments, a few tears, tears to linen folds, each section with a key map pasted to the marbled wrapper.

Dimensions
1840 by 1760mm (72.5 by 69.25 inches).

Scale
6 inches to 1 statute mile.

References
Hyde 91 (17).

First published in 1862, the map was to go through 21 revisions until its final appearance in 1901. "The plates of Stanford's Library Map of London' were used in the production of a number of other maps including the School Board Map of London, Booth's Descriptive Map of London Poverty, several metropolitan improvement maps, a map showing the distribution of troops lining the streets for the 1897 royal jubilee, another showing the routes taken by the infantry from their quarters to their places on the streets on the occasion of Edward VII's coronation, and Stanford's plan of the Metropolitan Borough of Poplar, 1903." (Hyde)

“The People’s Forest”

97 **WYLD, James**

Epping Forest as Awarded in Act [manuscript label].

Publication
[c. 1895].

Description
Lithographic map, printed in green and black, dissected and mounted onto linen, two folds with short tears.

Dimensions
1000 by 480mm (39.25 by 19 inches).

Maps of Epping Forest, one of the few ancient woodlands in the Southeast of Britain, are rather rare. The Epping Forest Act of 1878 was passed saving the forest from enclosure, and when Queen Victoria visited Chingford on 6 May 1882 she declared: “It gives me the greatest satisfaction to dedicate this beautiful forest to the use and enjoyment of my people for all time” and it thus became “The People’s Forest”.

Provenance: From the Commons & Footpath Preservation Society, who had campaigned successfully to stop the enclosure of Epping Forest, with their oval stamp on the verso. Pencil note in upper right-hand segment reading: “This plan is a reduction of the one signed by the arbitrator in 1882. The only additions since are Oak Hill Enclosure and Part of Higham Park containing 30 acres.”

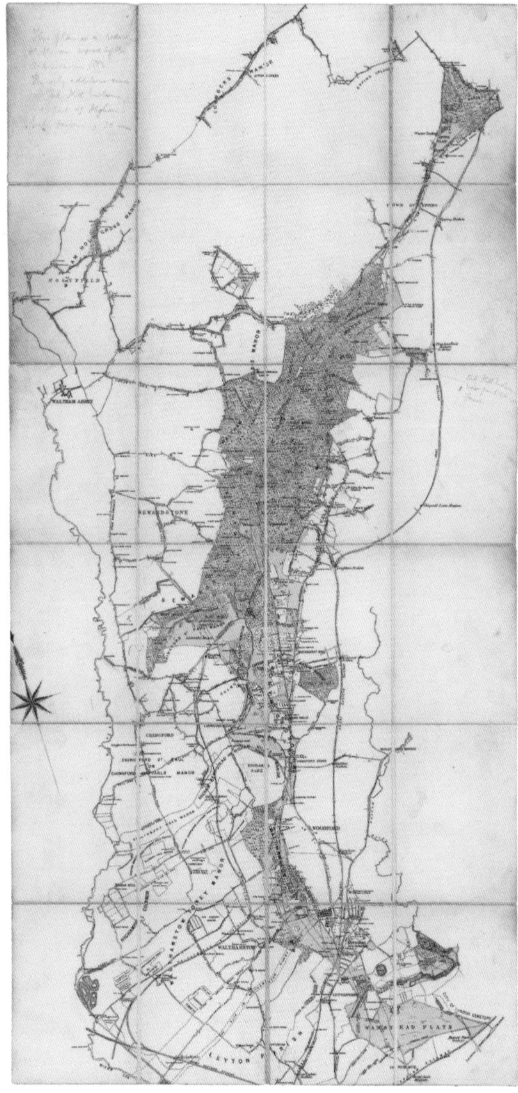

The first directly elected local government body for London

98 **LONDON COUNTY COUNCIL**

Municipal Map of London.

Publication
London, Martin, Hook & Larkin, Photolitho. Gt. Newport St. W.C. Estates & Valuation Department, 9 Spring Gardens, 1913.

Description
Chromolithograph, key sheet, and 28 sheets, each measuring 406 by 580 mm.

Dimensions
1900 by 2540mm (74.75 by 100 inches).

Scale
6 inches to 1 statute mile.

References
Hyde, pp.42–43.

An extremely detailed plan of London, based on earlier Ordnance Survey maps. Title in margin at top center on all sheets. Symbols for routes, parks etc. in margin top left, characteristics and symbols for boundaries in margin bottom left, scale-bar in margin bottom center, and symbols for public buildings in margin bottom right. Sheet 1 stamped: "Based upon the Ordnance Survey Map, with the sanction of the Controller of H.M. Stationery Office".

In 1894 the London County Council was given approval to produce a plan displaying all the freeholds in the newly created county, using 25-inch Ordnance plans as the base-map. It also used a variety of sources including plans in the possession of livery companies, paving trusts, docks, railways, and the Middlesex Register. The 'Ground Plan of London' as it was referred to, was finished in 1910. Alongside this Ground Plan was the 'Annual Map of London'. Unlike the Ground Plan, the Annual Map was printed and published as a thirty-five sheet map; however, its sale was restricted to county officials. The Annual Map was phased out when the 'Municipal Map of London' began publication. According to Hyde, the 'Municipal Map of London' this was first published in 1914; however, the present example is dated 1913.

The London County Council was created in 1889 as the first directly elected local government body for London. It had authority over most local divisions, such as public assistance, housing, health services, education and transport. Post-World War II, it was clear that the LCC could not cope with the demands being put on it, and it was replaced by the Greater London Council, which covered a larger area, and had greater resources.

First state of Beck's iconic tube map

99 **BECK, Harry**

Map of London's Underground Railways. A new design for an old map. We would welcome your comments. Please write to Publicity Manager, 55, Broadway, Westminster, S.W.1.

Publication
London Transport, 55, Broadway, Westinster, S.W.1, [January, 1933].

Description
Chromolithograph plan, title, list of places of interest and theatres to verso.

Dimensions
142 by 202mm (5.5 by 8 inches).

A fine example of Beck's iconic map of the London Underground System.

The map was designed by the 29 year-old engineer Harry Beck. Abandoning the restrictions of a geographically correct layout, the map actually constitutes a diagram of the network, showing relationships rather than distances to scale. By using only verticals, horizontals and diagonals, and adopting a clear colour scheme, Beck created a design classic, both easy to use and aesthetically appealing. After the positive public response to the limited trial run issued in 1932, the design was formally adopted in 1933, becoming an essential part of London Transport's campaign to project itself as a modern, rational and efficient system. The design remains in use to this day, having become essential to the comprehension of complex transport networks all over the world.

The present example is the first state of the map issued in January of 1933: the interchange stations are marked with a diamond; the Piccadilly Line is under construction between Enfield West and Cockfosters, due to be opened mid-summer of 1933.

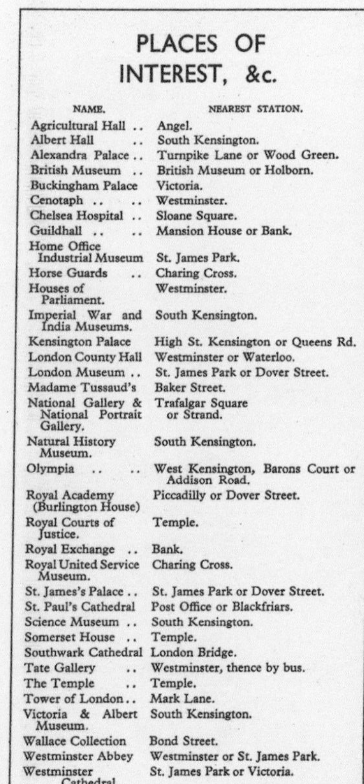

PLACES OF INTEREST, &c.

NAME.	NEAREST STATION.
Agricultural Hall ..	Angel.
Albert Hall ..	South Kensington.
Alexandra Palace ..	Turnpike Lane or Wood Green.
British Museum ..	British Museum or Holborn.
Buckingham Palace	Victoria.
Cenotaph	Westminster.
Chelsea Hospital ..	Sloane Square.
Guildhall	Mansion House or Bank.
Home Office Industrial Museum	St. James Park.
Horse Guards ..	Charing Cross.
Houses of Parliament.	Westminster.
Imperial War and India Museums.	South Kensington.
Kensington Palace	High St. Kensington or Queens Rd.
London County Hall	Westminster or Waterloo.
London Museum ..	St. James Park or Dover Street.
Madame Tussaud's	Baker Street.
National Gallery & National Portrait Gallery.	Trafalgar Square or Strand.
Natural History Museum.	South Kensington.
Olympia	West Kensington, Barons Court or Addison Road.
Royal Academy (Burlington House)	Piccadilly or Dover Street.
Royal Courts of Justice.	Temple.
Royal Exchange ..	Bank.
Royal United Service Museum.	Charing Cross.
St. James's Palace ..	St. James Park or Dover Street.
St. Paul's Cathedral	Post Office or Blackfriars.
Science Museum ..	South Kensington.
Somerset House ..	Temple.
Southwark Cathedral	London Bridge.
Tate Gallery ..	Westminster, thence by bus.
The Temple ..	Temple.
Tower of London ..	Mark Lane.
Victoria & Albert Museum.	South Kensington.
Wallace Collection	Bond Street.
Westminster Abbey	Westminster or St. James Park.
Westminster Cathedral.	St. James Park or Victoria.
Zoo	Camden Town or Chalk Farm.

(750M —1-33)

THEATRES.

THEATRE.	NEAREST STATION.
Adelphi	Charing Cross or Strand.
Aldwych	Aldwych or Temple.
Alhambra	Leicester Square.
Ambassadors ..	Leicester Square.
Apollo	Piccadilly.
Cambridge ..	Leicester Square.
Coliseum	Trafalgar Square or Strand.
Comedy	Piccadilly.
Court	Sloane Square.
Criterion	Piccadilly.
Daly's	Leicester Square or Piccadilly.
Dominion	Tottenham Court Road.
Drury Lane ..	Covent Garden or Temple.
Duchess	Aldwych or Temple.
Duke of York's ..	Leicester Square or Trafalgar Square.
Gaiety	Aldwych or Temple.
Garrick	Leicester Square or Trafalgar Square.
Globe	Piccadilly.
Haymarket ..	Piccadilly or Trafalgar Square.
Hippodrome ..	Leicester Square.
His Majesty's ..	Piccadilly or Trafalgar Square.
Holborn Empire ..	British Museum or Holborn.
Kingsway	British Museum or Holborn.
Leicester Square ..	Piccadilly or Leicester Square.
Little	Charing Cross or Strand.
London Pavilion ..	Piccadilly.
Lyceum	Covent Garden or Temple.
Lyric	Piccadilly.
New	Leicester Square or Strand.
Old Vic.	Waterloo.
Palace	Tottenham Ct. Rd. or Leicester Sq.
Palladium	Oxford Circus.
Phœnix	Tottenham Court Road.
Piccadilly	Piccadilly.
Playhouse	Charing Cross.
Prince Edward ..	Tottenham Court Rd. or Leicester Sq.
Prince's	British Museum or Holborn.
Prince of Wales ..	Piccadilly.
Queen's	Piccadilly.
Royal Opera House	Covent Garden.
Royalty	Tottenham Court Road.
St. James's ..	Dover Street.
St. Martin's ..	Leicester Square.
Sadlers Wells ..	Angel.
Saville	Tottenham Ct. Rd. or Leicester Sq.
Savoy	Charing Cross or Strand.
Scala	Goodge Street.
Shaftesbury ..	Piccadilly or Leicester Square.
Strand	Aldwych or Covent Garden.
Vaudeville	Charing Cross or Strand.
Victoria Palace ..	Victoria.
Westminster ..	Victoria.
Whitehall	Trafalgar Square or Charing Cross.
Winter Garden ..	Holborn or British Museum.
Wyndham's ..	Leicester Square.

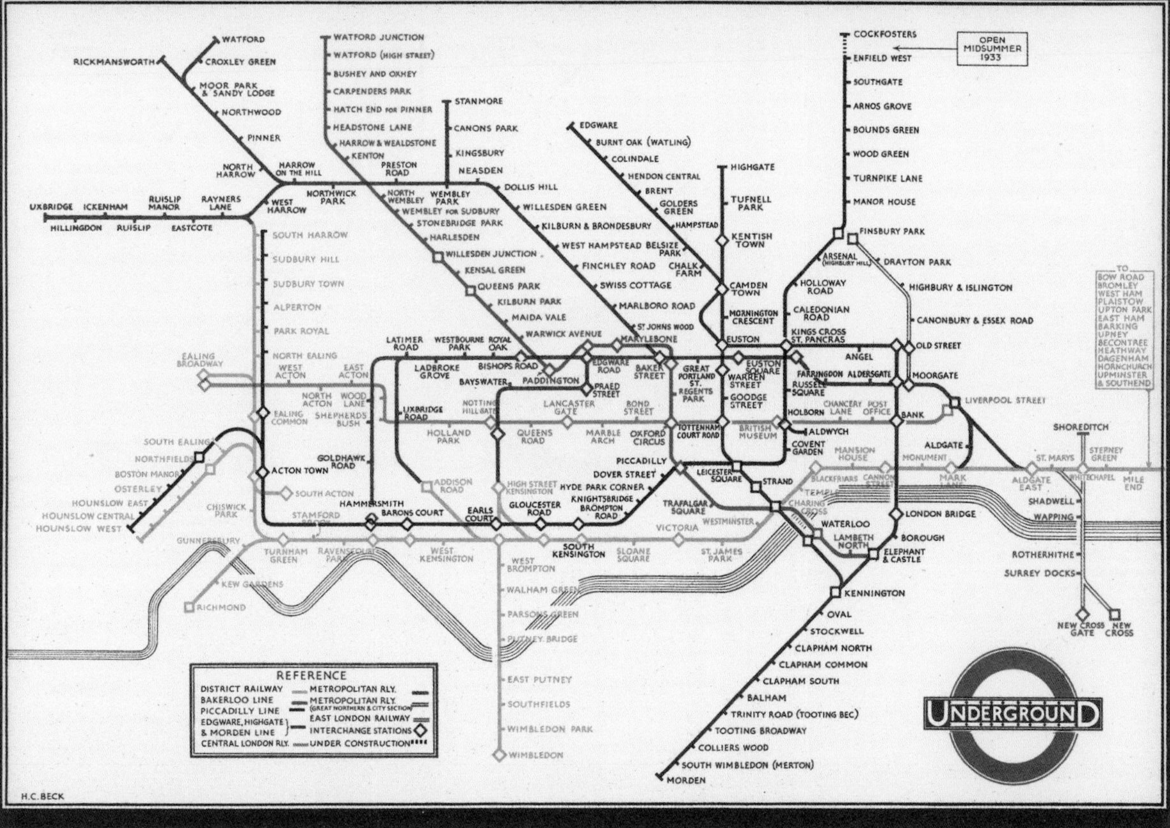
WATFORD
RICKMANSWORTH
CROXLEY GREEN
MOOR PARK & SANDY LODGE
NORTHWOOD
PINNER
NORTH HARROW
HARROW ON THE HILL
UXBRIDGE
ICKENHAM
RUISLIP MANOR
RAYNERS LANE
WEST HARROW
HILLINGDON
RUISLIP
EASTCOTE
NORTHWICK PARK
WATFORD JUNCTION
WATFORD (HIGH STREET)
BUSHEY AND OXHEY
CARPENDERS PARK
HATCH END FOR PINNER
HEADSTONE LANE
HARROW & WEALDSTONE
KENTON
PRESTON ROAD
NORTH WEMBLEY
WEMBLEY PARK
WEMBLEY FOR SUDBURY
STONEBRIDGE PARK
HARLESDEN
WILLESDEN JUNCTION
KENSAL GREEN
QUEENS PARK
KILBURN PARK
MAIDA VALE
WARWICK AVENUE
STANMORE
CANONS PARK
KINGSBURY
NEASDEN
DOLLIS HILL
WILLESDEN GREEN
KILBURN & BRONDESBURY
WEST HAMPSTEAD
FINCHLEY ROAD
SWISS COTTAGE
MARLBORO ROAD
ST JOHNS WOOD
EDGWARE
BURNT OAK (WATLING)
COLINDALE
HENDON CENTRAL
BRENT
GOLDERS GREEN
HAMPSTEAD
BELSIZE PARK
CHALK FARM
HIGHGATE
TUFNELL PARK
KENTISH TOWN
CAMDEN TOWN
MORNINGTON CRESCENT
EUSTON
COCKFOSTERS
OPEN MIDSUMMER 1933
ENFIELD WEST
SOUTHGATE
ARNOS GROVE
BOUNDS GREEN
WOOD GREEN
TURNPIKE LANE
MANOR HOUSE
FINSBURY PARK
ARSENAL (HIGHBURY HILL)
DRAYTON PARK
HOLLOWAY ROAD
HIGHBURY & ISLINGTON
CALEDONIAN ROAD
CANONBURY & ESSEX ROAD
KINGS CROSS ST. PANCRAS
OLD STREET
ANGEL
MOORGATE
TO
BOW ROAD
BROMLEY
WEST HAM
PLAISTOW
UPTON PARK
EAST HAM
BARKING
UPNEY
BECONTREE
HEATHWAY
DAGENHAM
HORNCHURCH
UPMINSTER
& SOUTHEND
SOUTH HARROW
SUDBURY HILL
SUDBURY TOWN
ALPERTON
PARK ROYAL
NORTH EALING
EALING BROADWAY
WEST ACTON
EAST ACTON
NORTH ACTON
WOOD LANE
EALING COMMON
SHEPHERDS BUSH
SOUTH EALING
NORTHFIELDS
BOSTON MANOR
OSTERLEY
HOUNSLOW EAST
HOUNSLOW CENTRAL
HOUNSLOW WEST
ACTON TOWN
SOUTH ACTON
CHISWICK PARK
GUNNERSBURY
KEW GARDENS
RICHMOND
STAMFORD BROOK
TURNHAM GREEN
RAVENSCOURT PARK
HAMMERSMITH
BARONS COURT
GOLDHAWK ROAD
WEST KENSINGTON
LATIMER ROAD
WESTBOURNE PARK
ROYAL OAK
LADBROKE GROVE
BISHOPS ROAD
PADDINGTON
BAYSWATER
UXBRIDGE ROAD
NOTTING HILL GATE
HOLLAND PARK
ADDISON ROAD
HIGH STREET KENSINGTON
EARLS COURT
GLOUCESTER ROAD
SOUTH KENSINGTON
WEST BROMPTON
WALHAM GREEN
PARSONS GREEN
PUTNEY BRIDGE
EAST PUTNEY
SOUTHFIELDS
WIMBLEDON PARK
WIMBLEDON
MARYLEBONE
EDGWARE ROAD
PRAED STREET
BAKER STREET
LANCASTER GATE
QUEENS ROAD
MARBLE ARCH
BOND STREET
OXFORD CIRCUS
GREAT PORTLAND ST.
REGENTS PARK
EUSTON SQUARE
WARREN STREET
GOODGE STREET
TOTTENHAM COURT ROAD
BRITISH MUSEUM
HOLBORN
ALDWYCH
COVENT GARDEN
RUSSELL SQUARE
FARRINGDON
ALDERSGATE
CHANCERY LANE
POST OFFICE
BANK
LIVERPOOL STREET
PICCADILLY
DOVER STREET
HYDE PARK CORNER
KNIGHTSBRIDGE
BROMPTON ROAD
LEICESTER SQUARE
STRAND
TRAFALGAR SQUARE
CHARING CROSS
WESTMINSTER
VICTORIA
SLOANE SQUARE
ST. JAMES PARK
BLACKFRIARS
TEMPLE
MANSION HOUSE
CANNON STREET
MONUMENT
MARK LANE
ALDGATE
ALDGATE EAST
ST. MARYS
SHOREDITCH
WHITECHAPEL
STEPNEY GREEN
MILE END
SHADWELL
WAPPING
ROTHERHITHE
SURREY DOCKS
NEW CROSS GATE
NEW CROSS
WATERLOO
LAMBETH NORTH
LONDON BRIDGE
BOROUGH
ELEPHANT & CASTLE
KENNINGTON
OVAL
STOCKWELL
CLAPHAM NORTH
CLAPHAM COMMON
CLAPHAM SOUTH
BALHAM
TRINITY ROAD (TOOTING BEC)
TOOTING BROADWAY
COLLIERS WOOD
SOUTH WIMBLEDON (MERTON)
MORDEN
REFERENCE
DISTRICT RAILWAY
BAKERLOO LINE
PICCADILLY LINE
EDGWARE, HIGHGATE & MORDEN LINE
CENTRAL LONDON RLY.
METROPOLITAN RLY.
METROPOLITAN RLY. (GREAT NORTHERN & CITY SECTION)
EAST LONDON RAILWAY
INTERCHANGE STATIONS
UNDER CONSTRUCTION
UNDERGROUND
H.C. BECK

The second state of a design classic

100 **BECK, Harry**

Underground Railways of London. London Underground Transport. Issued Free.

Publication
London Transport, 55, Broadway, Westminster, S.W.1, [June 1933].

Description
Chromolithograph plan, title, list of places of interest and theatres to verso.

Dimensions
142 by 202mm (5.5 by 8 inches).

An example of the second state of the map, issued in June of 1933. Minor changes have been made: the interchange stations are now marked with a circle; the Piccadilly Line now extends to Cockfosters; the Metropolitan Line to Aylesbury; and there is an escalator connection between Bank and Monument.

PLACES OF INTEREST

NAME	NEAREST STATION
Agricultural Hall ..	Angel
Albert Hall	South Kensington
Alexandra Palace ..	Turnpike Lane or Wood Green
British Museum ..	Holborn (Kingsway)
Buckingham Palace ..	Victoria
Cenotaph	Westminster
Chelsea Hospital ..	Sloane Square
Guildhall	Mansion House or Bank
Hampstead Heath ..	Hampstead
Hampton Court ..	Hammersmith, thence tram or Wimbledon, thence trolleybus
Home Office Industrial Museum	St. James Park
Horse Guards ..	Charing Cross
Houses of Parliament	Westminster
Imperial War and India Museums	South Kensington
Kensington Palace ..	High St. Kensington or Queens Road
Kew Gardens ..	Kew Gardens
London County Hall	Westminster or Waterloo
London Museum ..	St. James Park or Green Park
Lord's	St. John's Wood
Madame Tussaud's	Baker Street
National Gallery and National Portrait Gallery	Trafalgar Square or Strand
Natural History Museum	South Kensington
Olympia	West Kensington, Barons Court or Addison Road
Oval	Oval
Royal Academy (Burlington House)	Piccadilly or Green Park
Royal Courts of Justice	Temple
Royal Exchange ..	Bank
Royal United Service Museum	Charing Cross or Trafalgar Square
St. James's Palace ..	St. James Park or Green Park
St. Paul's Cathedral..	Post Office or Blackfriars
Science Museum ..	South Kensington
Somerset House ..	Temple
Southwark Cathedral	London Bridge
Tate Gallery.. ..	Westminster, thence by bus
The Temple	Temple
Tower of London ..	Mark Lane
Victoria and Albert Museum	South Kensington
Wallace Collection..	Bond Street or Baker Street
Wembley Stadium ..	Wembley Park or Sudbury Town, thence tram
Westminster Abbey	Westminster or St. James Park
Westminster Cathedral	St. James Park or Victoria
White City Stadium	Wood Lane
Zoo	Camden Town, Chalk Farm or St. Johns Wood

33—9636.

UNDERGROUND RAILWAYS OF LONDON

LONDON UNDERGROUND TRANSPORT

ISSUED FREE

LONDON TRANSPORT

55, Broadway, Westminster, S.W.1.

WATERLOW & SONS LIMITED, LONDON, DUNSTABLE & WATFORD.

THEATRES

THEATRE	NEAREST STATION
Adelphi ..	.. Charing Cross or Strand
Aldwych ..	.. Aldwych or Temple
Alhambra ..	.. Leicester Square
Ambassadors	.. Leicester Square
Apollo ..	.. Piccadilly
Cambridge ..	.. Leicester Square
Coliseum ..	.. Trafalgar Square or Strand
Comedy ..	.. Piccadilly
Criterion ..	.. Piccadilly
Daly's ..	.. Leicester Square or Piccadilly
Drury Lane ..	.. Covent Garden or Temple
Duchess ..	.. Aldwych or Temple
Duke of York's	.. Leicester Sq. or Trafalgar Sq.
Fortune ..	.. Covent Garden or Holborn (Kingsway)
Gaiety ..	.. Aldwych or Temple
Garrick ..	.. Leicester Sq. or Trafalgar Sq.
Globe ..	.. Piccadilly
Haymarket ..	.. Piccadilly or Trafalgar Square
Hippodrome	.. Leicester Square
His Majesty's	.. Piccadilly or Trafalgar Square
Holborn Empire	.. Holborn (Kingsway)
Kingsway ..	.. Holborn (Kingsway)
Little	.. Charing Cross or Strand
London Pavilion	.. Piccadilly
Lyceum ..	.. Covent Garden or Temple
Lyric	.. Piccadilly
New	.. Leicester Sq. or Trafalgar Sq.
Old Vic ..	.. Waterloo
Palace ..	.. Tottenham Court Road or Leicester Square
Palladium ..	.. Oxford Circus
Phœnix ..	.. Tottenham Court Road
Piccadilly ..	.. Piccadilly
Playhouse ..	.. Charing Cross
Prince Edward	.. Tottenham Court Road or Leicester Square
Prince's ..	.. Holborn (Kingsway)
Prince of Wales	.. Piccadilly
Queen's ..	.. Piccadilly
Royal Opera House..	Covent Garden
Royalty ..	.. Tottenham Court Road
St. James's ..	.. Green Park
St. Martin's ..	.. Leicester Square
Sadlers Wells	.. Angel
Saville ..	.. Tottenham Court Road or Leicester Square
Savoy.. ..	.. Charing Cross or Strand
Scala	.. Goodge Street
Shaftesbury ..	.. Piccadilly or Leicester Square
Shilling Theatre	.. Putney Bridge
Strand ..	.. Aldwych or Covent Garden
Vaudeville ..	.. Charing Cross or Strand
Victoria Palace	.. Victoria
Westminster	.. Victoria
Whitehall ..	.. Trafalgar Sq. or Charing Cross
Winter Garden	.. Holborn (Kingsway)
Wyndham's ..	.. Leicester Square

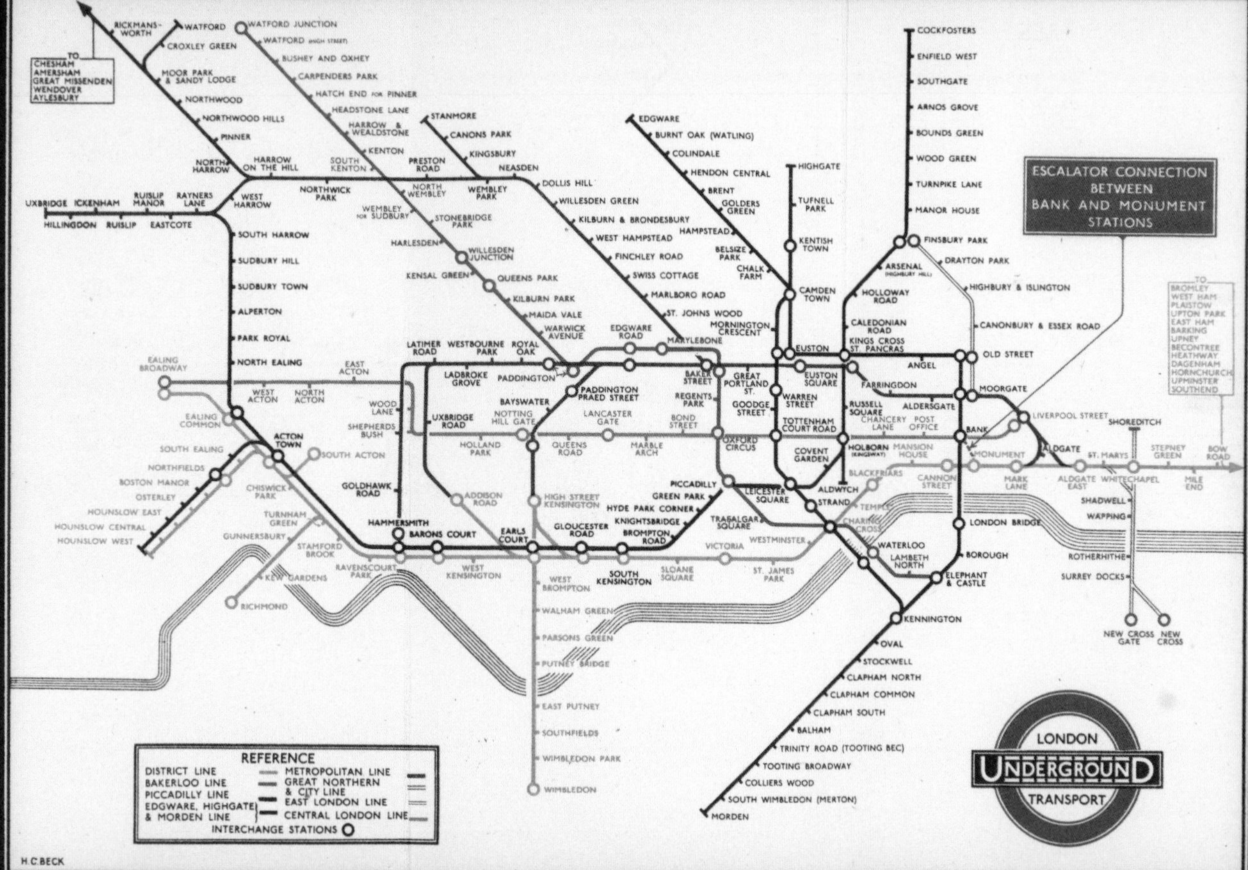
ESCALATOR CONNECTION BETWEEN BANK AND MONUMENT STATIONS
TO CHESHAM AMERSHAM GREAT MISSENDEN WENDOVER AYLESBURY
TO BROMLEY WEST HAM PLAISTOW UPTON PARK EAST HAM BARKING UPNEY BECONTREE HEATHWAY DAGENHAM HORNCHURCH UPMINSTER SOUTHEND
RICKMANSWORTH
WATFORD
CROXLEY GREEN
MOOR PARK & SANDY LODGE
NORTHWOOD
NORTHWOOD HILLS
PINNER
NORTH HARROW
HARROW ON THE HILL
WATFORD JUNCTION
WATFORD (HIGH STREET)
BUSHEY AND OXHEY
CARPENDERS PARK
HATCH END FOR PINNER
HEADSTONE LANE
HARROW & WEALDSTONE
KENTON
SOUTH KENTON
NORTH WEMBLEY
WEMBLEY FOR SUDBURY
STONEBRIDGE PARK
HARLESDEN
WILLESDEN JUNCTION
KENSAL GREEN
QUEENS PARK
KILBURN PARK
MAIDA VALE
WARWICK AVENUE
STANMORE
CANONS PARK
KINGSBURY
WEMBLEY PARK
PRESTON ROAD
NORTHWICK PARK
NEASDEN
DOLLIS HILL
WILLESDEN GREEN
KILBURN & BRONDESBURY
WEST HAMPSTEAD
FINCHLEY ROAD
SWISS COTTAGE
MARLBORO ROAD
ST. JOHNS WOOD
EDGWARE
BURNT OAK (WATLING)
COLINDALE
HENDON CENTRAL
BRENT
GOLDERS GREEN
HAMPSTEAD
BELSIZE PARK
CHALK FARM
HIGHGATE
TUFNELL PARK
KENTISH TOWN
CAMDEN TOWN
MORNINGTON CRESCENT
COCKFOSTERS
ENFIELD WEST
SOUTHGATE
ARNOS GROVE
BOUNDS GREEN
WOOD GREEN
TURNPIKE LANE
MANOR HOUSE
FINSBURY PARK
ARSENAL (HIGHBURY HILL)
DRAYTON PARK
HIGHBURY & ISLINGTON
HOLLOWAY ROAD
CALEDONIAN ROAD
CANONBURY & ESSEX ROAD
KINGS CROSS ST. PANCRAS
OLD STREET
ANGEL
UXBRIDGE
ICKENHAM
RUISLIP MANOR
RAYNERS LANE
WEST HARROW
HILLINGDON
RUISLIP
EASTCOTE
SOUTH HARROW
SUDBURY HILL
SUDBURY TOWN
ALPERTON
PARK ROYAL
NORTH EALING
EALING BROADWAY
EAST ACTON
WEST ACTON
NORTH ACTON
EALING COMMON
SOUTH EALING
NORTHFIELDS
BOSTON MANOR
OSTERLEY
HOUNSLOW EAST
HOUNSLOW CENTRAL
HOUNSLOW WEST
ACTON TOWN
SOUTH ACTON
CHISWICK PARK
TURNHAM GREEN
GUNNERSBURY
STAMFORD BROOK
KEW GARDENS
RICHMOND
RAVENSCOURT PARK
HAMMERSMITH
BARONS COURT
WEST KENSINGTON
LATIMER ROAD
WESTBOURNE PARK
ROYAL OAK
LADBROKE GROVE
PADDINGTON
PADDINGTON PRAED STREET
EDGWARE ROAD
MARYLEBONE
BAKER STREET
GREAT PORTLAND ST.
EUSTON
EUSTON SQUARE
REGENTS PARK
WARREN STREET
GOODGE STREET
WOOD LANE
SHEPHERDS BUSH
UXBRIDGE ROAD
BAYSWATER
NOTTING HILL GATE
LANCASTER GATE
BOND STREET
MARBLE ARCH
HOLLAND PARK
QUEENS ROAD
OXFORD CIRCUS
TOTTENHAM COURT ROAD
RUSSELL SQUARE
FARRINGDON
ALDERSGATE
CHANCERY LANE
POST OFFICE
MOORGATE
LIVERPOOL STREET
SHOREDITCH
BANK
MANSION HOUSE
MONUMENT
ALDGATE
ST. MARYS
STEPNEY GREEN
BOW ROAD
MILE END
ALDGATE EAST
WHITECHAPEL
MARK LANE
CANNON STREET
BLACKFRIARS
HOLBORN (KINGSWAY)
COVENT GARDEN
ALDWYCH
LEICESTER SQUARE
PICCADILLY
GREEN PARK
HYDE PARK CORNER
KNIGHTSBRIDGE
BROMPTON ROAD
GOLDHAWK ROAD
ADDISON ROAD
HIGH STREET KENSINGTON
EARLS COURT
GLOUCESTER ROAD
SOUTH KENSINGTON
SLOANE SQUARE
VICTORIA
ST. JAMES PARK
WESTMINSTER
TRAFALGAR SQUARE
STRAND
TEMPLE
CHARING CROSS
WATERLOO
LAMBETH NORTH
LONDON BRIDGE
BOROUGH
ELEPHANT & CASTLE
KENNINGTON
OVAL
STOCKWELL
CLAPHAM NORTH
CLAPHAM COMMON
CLAPHAM SOUTH
BALHAM
TRINITY ROAD (TOOTING BEC)
TOOTING BROADWAY
COLLIERS WOOD
SOUTH WIMBLEDON (MERTON)
MORDEN
WEST BROMPTON
WALHAM GREEN
PARSONS GREEN
PUTNEY BRIDGE
EAST PUTNEY
SOUTHFIELDS
WIMBLEDON PARK
WIMBLEDON
SHADWELL
WAPPING
ROTHERHITHE
SURREY DOCKS
NEW CROSS GATE
NEW CROSS
REFERENCE
DISTRICT LINE
BAKERLOO LINE
PICCADILLY LINE
EDGWARE, HIGHGATE & MORDEN LINE
METROPOLITAN LINE
GREAT NORTHERN & CITY LINE
EAST LONDON LINE
CENTRAL LONDON LINE
INTERCHANGE STATIONS
LONDON
UNDERGROUND
TRANSPORT
H.C.BECK

Select Bibliography

Barker, Felix & Jackson, Peter, *The History of London in Maps*, London, Guild Publishing, 1990.

Chubb, Thomas, *The Printed Maps in the Atlases of Great Britain and Ireland: A Bibliography. 1579–1870*, 2nd Ed., London, Burrow & Co., 1966.

Glanville, Philippa, *London in Maps*, London, The Connoisseur, 1972.

Howgego, James, *Printed Maps of London circa 1553–1850*, 2nd Ed., London, Dawson & Sons Ltd., 1978.

Hyde, Ralph, *Printed Maps of Victorian London, 1851–1900*, London, Dawson & Sons Ltd., 1975.

Hyde, Ralph, *Ogilby and Morgan's Survey of the City of London, 1676*. Notes to accompany a reproduction by Harry Margary, 1976.

Krogt, Peter van der, *Koeman's Atlantes Neerlandici*, Houten, Vol. I–IV, Hes & de Graaf, 1997–2010.

Pennington, Richard, *A descriptive catalogue of the etched work of Wenceslaus Hollar 1607–1677*, Cambridge, CUP, 1982.

Rodger, E.M., *The large-scale county maps of the British Isles 1596–1850*, Oxford, OUP, 1972.

Skelton, R.A., *County Maps of the British Isles 1579-1850*, London, Carta Press, 1970.

Worms, Lawrence & Baynton-Williams, Ashley, *British Map Engravers*, London, Rare Book Society, 2011.